WHEAT CULTIVATOR

WHEAT CULTIVATOR

P C SINGH
SANU XAVIER

KNOWLEDGE BAKERS

KNOWLEDGE BAKERS

ISBN: 978-93-49964-03-7

Published by : **Knowledge Bakers**
Pune, India
info@knowledgebakers.com

PREFACE

Wheat Cultivator serves as a comprehensive guide dedicated to the cultivation of one of the world's most important staple crops. This book is designed for farmers, agricultural students, agronomists, and professionals involved in the wheat production industry, offering detailed insights into the techniques, technologies, and practices that lead to the successful cultivation of wheat. The primary aim is to equip readers with the knowledge needed to improve yield, quality, and sustainability in wheat farming, addressing both traditional methods and contemporary innovations.

The book begins by providing an overview of wheat as a crop, covering its history, importance in global food security, and its role in diverse agricultural systems. It discusses the various species of wheat, including common wheat (Triticum aestivum) and durum wheat (Triticum durum), and their suitability for different climates and soils. The introductory chapters emphasize the key aspects of wheat cultivation, including selecting the right varieties, understanding soil health, and assessing climatic conditions necessary for optimal growth.

As wheat is susceptible to numerous pests, diseases, and weeds, the book highlights integrated pest management (IPM) strategies, focusing on the prevention and control of common wheat diseases such as rusts, fusarium, and blights. It discusses various pest management techniques, including biological control, chemical control, and cultural practices, helping farmers mitigate the risks of yield loss due to pests and pathogens. Weed control practices, such as pre- and post-emergence herbicide application and mechanical control, are also covered in detail to assist farmers in managing weed competition effectively.

Water management plays a pivotal role in wheat cultivation, and Wheat Cultivator emphasizes the importance of proper irrigation techniques to optimize water use efficiency and reduce water wastage. The book discusses different irrigation systems, including surface irrigation, drip irrigation, and sprinkler systems, while considering factors such as climate, soil type, and crop water requirements. It also addresses the challenges of water scarcity in many wheat-growing regions and explores water conservation strategies that help farmers achieve high yields with minimal water usage.

Harvesting and post-harvest management are key components of wheat production, and the book dedicates considerable attention to these areas. It provides a step-by-step guide to harvesting wheat at the right time, ensuring the grain reaches peak maturity for optimal quality and storage life. The book also addresses post-harvest handling techniques, including threshing, cleaning, and storage methods that preserve grain quality and minimize losses due to pests, diseases, or poor storage conditions.

Furthermore, the book discusses the role of wheat cultivators in the broader agricultural value chain, exploring the economic aspects of wheat production. Topics include market access, price fluctuations, and the economics of wheat farming, with a focus on how farmers can improve profitability while managing risks. It also addresses the importance of supporting smallholder farmers in developing countries, providing practical advice on financing options, crop insurance, and access to agricultural extension services.

The book concludes by discussing emerging trends in wheat cultivation, such as genetic improvements through breeding programs and the use of genetically modified wheat to enhance resistance to disease, pests, and environmental stress. It also explores the potential impact of climate change on wheat production, offering strategies for adapting to changing weather patterns and ensuring long-term food security.

Wheat Cultivator is an essential reference for anyone involved in the cultivation of wheat, offering both practical knowledge and scientific insights that contribute to better farming practices and enhanced crop production. Whether you are a novice farmer looking to start wheat cultivation or an experienced agronomist seeking to optimize your wheat-growing techniques, this book provides the tools and knowledge necessary to succeed in the dynamic world of wheat farming.

P C Singh

Sanu Xavier

CONTENTS

Chapter 1

SELECTION OF VARIETIES FOR WHEAT CULTIVATION

1.1 INTRODUCTION

Wheat is a triticum species which is a crop that is grown in diversified environments. It is a staple food of millions of people. Around one-sixth of the total arable land in the world is cultivated with wheat. Whereas paddy is mainly cultivated in Asia, wheat is grown in all the continents of the world. It supplies about 20 per cent of the food calories for the world's growing population. Global wheat production touched 622.2 million tones in 2007-08. India is the second largest producer of wheat after China. Wheat has a distinct place among the food grain crops. Carbohydrate and protein are two main constituents of wheat. On average wheat contains 11-12% protein.

- ☆ Scientific name : *Triticum aestivum* L.
- ☆ Local name : Gehun
- ☆ Hindi name : Gehun
- ☆ Crop season : Rabi

Wheat is grown mainly in two seasons across the world; winter and spring. Winter wheat is grown in cold countries like Europe, U.S.A., Australia, Russia Federation etc. while spring wheat is grown in Asia and a part of U.S.A. Spring wheat matures in 120-130 days while winter wheat takes 240-300 days for maturity. Due to this reason productivity of winter wheat is higher in comparison to spring wheat. Considering the quality wheat has been divided into two categories:

1. soft wheat
2. hard wheat

Triticum aestivum (bread wheat) is known as soft wheat and Triticum durum is known as hard wheat. Wheat is generally grown intended for food for humans, but lesser quality wheat and the nutrient-dense by-products of flour refining are used for animal feed. Wheat is also used for whiskey and beer production, and the husk can be separated and ground into bran. Before the introduction of corn into Europe, wheat was the principal source of starch for sizing paper and cloth.

1.1.1 Wheat Classes

There are six major classes of wheat:

Hard Wheat

- ☆ Hard red winter wheat is a high protein wheat used mostly for breads and all-purpose flour, and as an adjunct in other flours to increase protein content. This class of wheat accounts for more than 40% of the U.S. wheat crop and half of U.S. wheat exports.
- ☆ Durum wheat is a spring wheat, and may be either white or red. It is the hardest of all U.S. wheats, offers both high protein and high gluten content, and used to make the semolina flour used for premium pasta products and some Mediterranean breads.
- ☆ Hard red spring wheat is the highest protein wheat, and is used for bread, hard baked goods, all-purpose flour, and flour blends.
- ☆ Hard white wheat is medium protein grain that is closely related to red wheat except for color, in milling and baking qualities. However, it offers a milder, sweeter flavor, and is used in yeast breads, hard rolls, bulgur, tortillas, oriental noodles, whole wheat and all-purpose flowers, and used in brewing. It is the newest class of wheat to be grown in the United States.

Soft Wheat

- ☆ Soft red winter wheat has a low to medium protein content, and is used for breads and blending. It is used to make cookies, cakes, donuts, and other fine pastries as well as flat breads, and crackers.
- ☆ Soft white wheat is a low protein wheat, but offers high yields to growers. It provides a whiter product for high quality cakes, crackers, cookies, pastries, and Asian-style noodles (bakery products other than breads), and is ideally suited to Middle Eastern flatbreads.

1.1.2 Classification of wheat

Wheat is an annual plant of Gramineae family. It belongs to genus Triticum. Although as many as 18 species of wheat have been recognized but only few are of importance in agriculture.

Wheat can be classified based on:

- ☆ Colour (e.g., red, yellow, white)
- ☆ Planting season: spring wheat, winter wheat
- ☆ Characteristics of grain: durum, hard bread wheat and soft wheat

Classification of wheat grains has significant importance in determining the market value of wheat varieties. Quality of final products depends on the use of specific wheat varieties. Wheat class identification is also necessary for plant breeders to predict yield and quality.

The morphological features of grains are heritable characters and play an important role in wheat variety recognition. In the current grain grading systems, grain type and quality are inspected through visual assessment by experts and trained inspectors which is subjective, and tedious. Computer vision system offers an objective and quantitative method for estimation of morphological parameters and quality of agricultural products to obtain quick and more reliable results.

In grain classification, several techniques like statistical, artificial neural networks and fuzzy logic have been used. The classification performances of different neural network topology were compared by using morphological features of Canada Western Amber Durum (CWAD) wheat, Canada Western Red Spring (CWRS) wheat, oats, rye and barley.

Wheat classification can be done based on following steps:

- ☆ sample preparation
- ☆ image acquisition
- ☆ pre-processing operations
- ☆ feature extraction
- ☆ feature selection
- ☆ classify wheat cultivars

Wheat cultivars classification steps are explained in following:

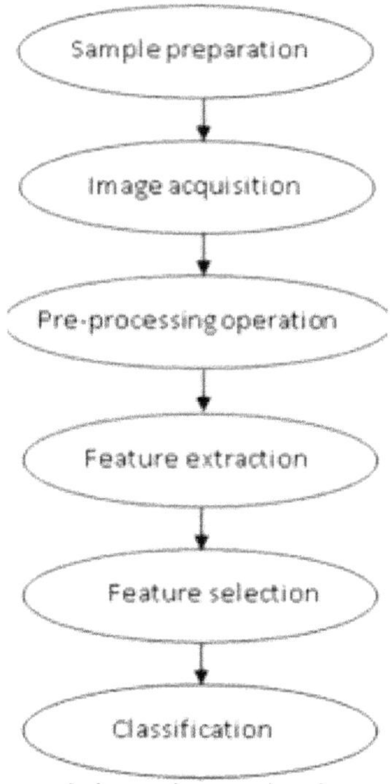

In India there exists only three species of wheat:

- ☆ Triticum aestivum (common bread wheat),
- ☆ Triticum durum (macaroni or durum wheat)
- ☆ Triticum dicoccum (emmer wheat)

The bread wheat (Triticum aestivum) is the most important species accounting to a little over 87 per cent of the total wheat production of India followed by the durum wheat (about 12 per cent) and dicoccum wheat (about 1 per cent). The common wheat (T. aestivum) which is good for chapati making and bakery products is grown in whole of North India and also in Central and South India. The durum wheat which is good for 'suji', 'semya', 'sphagetti' and 'macaronil' is grown only in Central and South India and that too under rainfed conditions. The dicoccum wheat is grown only on limited acreage in Tamil Nadu, Andhra Pradesh, Maharashtra and Gujarat. It is good for the South India dish 'Uppumav'.

1.1.3 Status of wheat

Wheat is one of the most important crops to humankind as it is a staple of many diets around the world. As per FAOSTAT, China produces more wheat than any other country, followed by India, Russia, and United States.

Germany (24,481,600 Tonnes)

- ☆ Germany is one of the largest grain producers in the European Union, and as such its farming practices are monitored by E.U. market management and affected by E.U. rules and regulations.
- ☆ According to estimates, the flour mills of Germany process 24 million tonnes of wheat each year.
- ☆ Winter wheat is cultivated throughout the country, though concentrated primarily in central Germany.
- ☆ Bayren (19%) and Niedersachsen (17%) are largest wheat producing states in country.
- ☆ If normal climatic conditions prevail, the crop is planted in October and harvested in August the following year.

Ukraine (26,208,980 Tonnes)

- ☆ Wheat is grown all across the Ukraine, with majority of yields contributed by central and south-central regions of country.
- ☆ The crop is planted in fall season and harvested between July and August the following year.
- ☆ Known as "breadbasket of Europe", Ukraine predominately produces hard, red variant of winter wheat, used in breadmaking.

Pakistan (26,674,000 Tonnes)

- ☆ Wheat in Pakistan acts as the leading food grain, and a staple food in diets of Pakistani people.
- ☆ The area cultivated from wheat cultivation increased by 4.4%.
- ☆ Wheat is cultivated in all parts of Pakistan, with maximum yields being obtained from Punjab and Sindh.
- ☆ The alluvial deposits by Indus River contributes to fertility of soil in such areas, and encourages the cultivation of cereal crops, including wheat, on a massive scale.

Canada (29,984,200 Tonnes)

- ☆ Wheat is the most important crop grown in Canada, and several varieties of wheat are cultivated which covers winter wheat, dark northern spring wheat and durum wheat.
- ☆ The wheat grains are used for several purposes, for generating flour for bakery industry and feed for livestock.

- ☆ Saskatchewan is the topmost DNA and durum wheat producing region in Canada, followed by Alberta.
- ☆ Ontario, accounts for 82% of total winter wheat production in Canada.

Australia (31,818,744 Tonnes)

- ☆ Wheat is one of the major winter crops grown in Australia.
- ☆ Western Australia, Victoria, New South Wales, and Queensland are the top wheat producing states of the continent.
- ☆ The crop is sown in autumn months and harvested in spring or winter, depending on the environmental conditions.
- ☆ Western Australia is largest exporter of wheat, especially to those countries in Asia and Middle East, generating earnings of around $2 billion annually from such sales.
- ☆ For past 30 years, there has been a constant 1% annual rise in Western Australia's wheat production.
- ☆ The east coast of Australia, produces wheat for domestic consumption and feedstock.

France (36,924,938 Tonnes)

- ☆ France is the largest producer of wheat in Europe, with this crop being cultivated throughout the country.
- ☆ It is cultivated to the greatest extent in the northern regions of France.
- ☆ The French state of Centre is the leading wheat producing region in the country followed by Picardie.
- ☆ Winter wheat is the main variety of wheat grown in this country, and is planted in the autumn season and harvested in August of the following year.

United States (47,370,880 Tonnes)

- ☆ Wheat, the principal cereal grain of the United States, is grown all over the country.
- ☆ Around 47 million tonnes of wheat are produced in USA, which ranks fourth in the world in terms of quantity of wheat produced.
- ☆ As per USDA, there are eight varieties of wheat grown in the country such as durum wheat, hard red winter wheat, hard red spring wheat, soft white wheat and hard white wheat.
- ☆ 70-80% of wheat produced in USA belongs to the category of winter wheat.

- ☆ Per USDA harvest data, North Dakota, Kansas and Montana were the largest producers of wheat in this country.

Russia (85,863,132 Tonnes)

- ☆ Russia is the third largest wheat producer in the world and was among the top five wheat exporting countries in the world.
- ☆ Winter wheat is the primary variety of wheat grown in the country.
- ☆ The crop is mostly raised in the western parts of Russia surrounding Moscow.
- ☆ The crop is planted between August and first week of October, and harvested between July and August the following year.

India (98,510,000 Tonnes)

- ☆ Wheat is the second most important cultivated food crop in India, and feeds hundreds of millions of Indians on a daily basis.
- ☆ It is an especially important staple food in the northern and northwestern states of the country.
- ☆ India accounts for about 8.7% of the total wheat production in the world, and 13% of all cultivated land in India is dedicated to cultivation of this crop.
- ☆ Uttar Pradesh, Punjab, Haryana, and Madhya Pradesh are the leading wheat producing states in the country.

China (134,340,630 tonnes)

- ☆ It is world's largest producer of wheat.
- ☆ It plays an important role in shaping grain market dynamics across the world.
- ☆ Around 126 million metric tons of wheat are produced by the country per year, on a land area of 24 million hectares.
- ☆ Wheat serves as one of the staple foods of the Chinese population, accounting for 40% of grain consumption of the country.
- ☆ Wheat is cultivated extensively in the Yellow River and Huai River Valleys of China, where the crop is rotated with maize.

Wheat Producing Countries

Country	Wheat Produced (Tonnes)
China	134,340,630
India	98,510,000
Russian Federation	85,863,132

United States of America	47,370,880
France	36,924,938
Australia	31,818,744
Canada	29,984,200
Pakistan	26,674,000
Ukraine	26,208,980
Germany	24,481,600

1.2 IMPORTANCE OF DIFFERENT TYPES OF WHEAT AND THEIR STATUS

1.2.1 Different types of Wheat

Wheat is classified as per growing seasons, hardness (hard/soft), color (red/white) and shape such as:

As per Seasons:

Winter Wheat

- It flower only after several weeks of cold weather.
- They are sown in autumn and harvested the following spring or summer in regions having cold winters.

Spring Wheat

- They are planted in the spring
- These wheat have no temperature requirement for flowering.
- They are sown in autumn in regions that are relatively warm year-round.

As per hardness, colour and shape:

Hard Red Winter

- ☆ It is the largest class of wheat produced each year.
- ☆ Most of the wheat of this class is grown south of Northern plains
- ☆ It is the main wheat class used to produce white sliced pan bread and variety of other yeast-leavened baked goods.
- ☆ It has wide range of protein content.

Hard Red Spring

- ☆ It contains highest protein content of all wheat classes between 13.5-14.5%.
- ☆ It has superior milling and baking properties and is used to produce bread products requiring strong gluten.
- ☆ It is blended with lower protein flours for improving bread making qualities.

Soft Red Winter

- ☆ It is not grown in significant quantities in northern Plains.
- ☆ In this, level and quality of protein is same as that in soft white wheat's.
- ☆ It is good for cakes, cookies, flat breads, pastries and crackers.

Durum

- ☆ It is milled into granular product called semolina, that is used for pasta products.

Hard White Wheat

- ☆ Developed from hard red wheat by removing genes for bran color keeping other desireable characteristics of red wheat.
- ☆ white wheat has no major genes for bran color.

Soft White Wheat

- ☆ It is mainly grown in Pacific Northwest.
- ☆ It is used in cakes, crackers, cookies, pastries, quick breads, muffins and snack foods.

1.2.2 Wheat varieties in India

There are many varieties of wheat available in India such as:

- ☆ PBW 1 Zn:
- ☆ The plant of this variety attains the height of 103 cm.

- ☆ The crop is ready to harvest in 151 days.
- ☆ It gives an average of 22.5 qtl/acre of crop yield.

UNNAT PBW 343:

- ☆ Good for irrigated and timely sown areas.
- ☆ It is harvested after 155 days.
- ☆ It is resistant to lodging, water logging conditions, karnal bunt and tolerant to blight.
- ☆ It gives average yield of 23.2 qtl/acre.

WH 542:

- ☆ It is good for timely sown, irrigated areas.
- ☆ It is ready for harvesting in 135-145 days.
- ☆ It is resistant to stripe rust, leaf rust and karnal bunt.
- ☆ It give average yield of 20 qtl/acre.

PBW 725:

- ☆ It is dwarf variety good for timely sown irrigated areas.
- ☆ It is resistant to yellow and brown rust.
- ☆ Its grains are amber, hard and medium bold.
- ☆ It is ready to harvest in 155 days.
- ☆ It gives average yield of 23 qtl/acre.

PBW 677:

- ☆ It is ready for harvest in 160 days.
- ☆ It gives average yield of 22.4 qtl/acre.

HD 2851:

- ☆ It is good for timely sowing and is grown in irrigated areas.
- ☆ It matures in 126-134 days
- ☆ The plant attains the height of 80-90cm.

WHD-912:

- ☆ It is a double dwarf Durum variety used for bakery in industry.
- ☆ It has protein content of 12%.
- ☆ It is resistant to yellow & brown which rust and karnal bunt.
- ☆ The yield is near about 21 qtl/ acre.

HD 3043:

- ☆ It gives average yield of 17.8 qtl/acre.
- ☆ It is high resistant against stripe rusts and leaf rust.
- ☆ It has higher value of bread loaf volume (cc), bread quality score.

WH 1105:

- ☆ It is a double dwarf variety with average plant height of 97 cm.
- ☆ Its grains are amber, hard, medium bold and lustrous.
- ☆ It is resistant to yellow rust and brown rust but susceptible to karnal bunt and loose smut diseases.
- ☆ It matures in 157 days and its average grain yield is 23.1 quintals per acre.

PBW 660:

- ☆ It is a dwarf variety with average plant height of 100 cm.
- ☆ Its grains are amber, hard, bold and lustrous with very good chapatti quality.
- ☆ It is resistant to yellow and brown rusts but susceptible to lose smut disease.
- ☆ It matures in about 162 days.
- ☆ Its average grain yield is 17.1 quintals per acre.

PBW-502:

- ☆ It is good for timely sown irrigated conditions.
- ☆ It is resistant to leaf rust and stripe rust.

HD 3086(PusaGautam):

- ☆ It gives average yield of 23 qtl/acre.
- ☆ It is resistant to yellow rust and brown rust.
- ☆ It meets all the criteria for superior bread making qualities.

HD 2967:

- ☆ It is double dwarf variety with an average plant height of 101 cm.
- ☆ In this, ears are medium dense.
- ☆ It is resistant to yellow and brown rust but susceptible to karnal bunt and loose smut diseases.
- ☆ It takes about 157 days to mature.
- ☆ Its yield is 21.5 qtl/acre.

DBW17:

- ☆ The height of the plant is 95 cm.
- ☆ In this, grains are amber hard, medium bold and lustrous.
- ☆ It is susceptible to new races of yellow rust and moderately resistant to brown rust.
- ☆ It matures in 155 days.
- ☆ The average yield is 23 qtl/acre.

PBW 621:

- ☆ It gets ready for harvesting within 158 days.
- ☆ It is resistant to yellow and brown rusts diseases.
- ☆ It has an average height of 100cm.

UNNAT PBW 550:

- ☆ It gets ready for harvesting within 145 days.
- ☆ It is resistant to yellow and brown rusts diseases.
- ☆ It has an average height of 86 cm.
- ☆ It gives an average yield of 23 qtl/acre.

TL 2908:

- ☆ It gets ready for harvesting within 153 days.
- ☆ It is resistant to mostly all major diseases.
- ☆ It has an average height of 113 cm.

PBW 175:

- ☆ It gets ready for harvesting within 165 days.
- ☆ It is resistant to Rust and Karnal bunt diseases.
- ☆ It has an average height of 110 cm.

PBW 527:

- ☆ It gets ready for harvesting within 160 days.
- ☆ It is resistant to yellow and brown rust diseases.
- ☆ It has an average height of 100 cm.

WHD 943:

- ☆ It gets ready for harvesting within 154 days.
- ☆ It is resistant to yellow and brown rust diseases.
- ☆ It has an average height of 93 cm.

PDW 291:

- ☆ It gets ready for harvesting within 155 days.
- ☆ It is resistant to yellow and brown rust, loose smut and flag smut diseases.
- ☆ It has an average height of 83 cm.

PDW 233:

- ☆ It gets ready for harvesting within 150 days.
- ☆ It is resistant to yellow and brown rust, loose smut and karnal bunt diseases.
- ☆ It has an average height of 98 cm.

PBW 590:

- ☆ It gets ready for harvesting within 128 days.
- ☆ It is resistant to yellow and brown rusts diseases.
- ☆ It has an average height of 80cm.

PBW 509:

- ☆ It gets ready for harvesting within 130 days.
- ☆ It is resistant to yellow and brown rusts diseases.
- ☆ It has an average height of 85cm.

PBW 373:

- ☆ It gets ready for harvesting within 140 days.
- ☆ It is resistant to brown rusts diseases.
- ☆ It has an average height of 90 cm.

RAJ-3765:

- ☆ It matures in 120-125 days.
- ☆ Heat tolerant and suitable for zero tillage, Susceptible to brown rust, moderately susceptible to stripe rust and karnal bunt.
- ☆ The yield is near about 21 qtl/ acre.

UP-2338:

- ☆ It matures in 125-130 days.
- ☆ It is susceptible to leaf rust and moderately susceptible to stripe rust.
- ☆ Susceptible to karnal bunt and tolerant to blight.
- ☆ The yield is near about 21 qtl/acre.

UP-2328:

- ☆ It matures in 130-135 days.
- ☆ Ear heads are hard, sarbati colour & medium size grains.
- ☆ It is suitable for irrigated areas.
- ☆ The yield is near about 20-22 qtl/acre.

Sonalika:

- ☆ It has early maturing single dwarf wheat with wide adaptation and attractive amber grains.
- ☆ It is suitable for late sowing and resistant to rusts.

Kalyansona:

- ☆ A double dwarf wheat with wide adaptation recommended for cultivation all over India.
- ☆ It is very vulnerable to rust.
- ☆ It grows only in rust free regions.

UP-(368):

- ☆ It is resistant to rust and karnal bunt.

WL-(711):

- ☆ It is single dwarf, high yielding and medium maturing variety.
- ☆ It is moderately susceptible to powdery mildew and karnal bunt.

UP-(319):

- ☆ It is triple dwarf wheat with high level of rust resistance.
- ☆ To avoid losses to shattering, it should be harvested at appropriate time.

1.3 SELECTION OF VARIETIES

Selecting many varieties serves as countless advantages as wheat growers can use different tools and publications to study each variety's strengths and weaknesses, and thereby selecting required wheat varieties which best suited as per their needs

Having many varieties to choose from gives wheat cultivator the luxury of being able to avoid planting of given variety due to very specific concerns that may have arisen in certain locations in previous years. When selecting a variety that is best suited for farming operation, wheat cultivator needs to consider certain traits which include:

- ☆ winter hardiness,
- ☆ disease resistance,
- ☆ yield potential,
- ☆ market opportunities,

- ☆ lodging resistance
- ☆ **Winter hardiness**

The winter hardiness ratings of most winter wheat varieties are good to excellent. Wheat cultivator who farm in areas should be particularly vigilant in selecting a variety which could be good for winter hardiness. Also, it is recommended that seeding practices are compromised, such as late seeding or seeding into inadequate stubble, will lead to winter hardiness of variety into criticality of crop's success.

Disease resistance

Resistance to diseases common to the area and/or farming practices is another important consideration when deciding on wheat variety. If wheat cultivator are in an area that commonly struggles with rusts, there are varieties that provide resistance to stem, stripe, and leaf rust. If wheat streak mosaic virus is a concern due to past occurrence or tight wheat rotations, there are other varieties available with resistance to the wheat curl mite, the vector of this disease. Producers in areas where Fusarium Head Blight (FHB) is an issue should also be aware that, although winter wheat can avoid this disease in many years, there are varieties that offer genetic resistance.

Yield potential

Yield potential of each variety is influenced by management practices and growing regions. Wheat yields are subject to risk. The genotype-environment interaction describes how well each variety of wheat seeds will respond to different growing conditions.

1.4 IDENTIFICATION OF SOURCES OF QUALITY SEED

Seed is the basic and most critical input for sustainable agriculture. Quality seeds are the most important farming inputs to achieve high yields and good produce can be achieved only when the seeds are used wisely. For wheat cultivator, access to affordable and high quality seeds is therefore important for increasing agricultural production and productivity. In many developing countries, the formal seed system has shifted from a publicly dominated sector to a privately dominated sector.

Quality seeds is another input necessary for agricultural productivity, and good quality seeds account for 20%-25% of increased crop productivity.

Agricultural seeds are produced by various agencies such as Indian Council of Agricultural Research and its research institutions, state agricultural universities, and national and state seeds corporations.

Wheat cultivator can take wheat seeds from various government, semi-government and private agencies such as:

Government agencies

- National Seeds Corporation: It produces and market commercial seeds.
- States Farms Corporation of India Limited: It produces breeder, foundation and certified seeds of high yielding varieties.
- **Private seed firms**
- Private seed firms emerges during 60's and 70's, which involves in breeding of seeds and releases inbred improved cultivars.
- They involve in supplying some seeds to respective seed cultivators.

1.5 PURCHASE OF VARIETY

1.5.1 Variety selection

- Select the type of wheat seed variety as per the need, yield, place, temperature.
- Before purchasing the type of seeds, compare the price with different vendors.
- Check the new variety for better or equivalent yield and disease traits by giving diversity or risk mitigation that suits the current market needs.
- Review the performance of varieties using testing procedure and information from supplier.

1.5.2 Purchasing variety

- Before purchasing different seed variety, wheat cultivator should make sure that the variety what he/she is purchasing should be free from any disease.
- Diversify the wheat varieties as well as crop type.
- Check the type of quality required as per the yield.
- Identify the price as per bulk purchase.
- Compare the price from various agencies before buying.

1.6 STORING VARIETY AS PER RECOMMENDED PROCEDURE IF THERE IS TIME FOR SOWING

1.6.1 Storage

- ☆ The grains should be thoroughly dried before storage.
- ☆ If seed is dry and cool, fungal activity and insect activity are low.
- ☆ Select a seed storage site that is cool and dry.
- ☆ The storage life of the grain is closely related to its moisture content.
- ☆ Grains with less than 10% moisture store well.
- ☆ The storage pits, bins or godowns should be moisture-proof and should be fumigated to keep down the stored grain pests.
- ☆ Wheat seed is storable for medium to long periods if kept under safe storage conditions.
- ☆ For wheat, high seed moisture is most damaging, and seed must be kept as dry as possible in storage.
- ☆ The response of wheat seed to high atmospheric humidity in storage varies with temperature.

1.6.2 Sowing time

- ☆ Wheat must be sown at the optimum time.
- ☆ Delayed sowing causes a gradual decline in the yield of wheat.
- ☆ The time of sowing is from 25 October to November:
    - For indigenous wheat: last week of October-long dwarf varieties; Kalyansona, Arjun, etc.
    - For first fortnight of November and for short duration dwarf wheats.
    - For second fortnight: best sowing time.
    - For late sown: By 1st week of December.
- ☆ For normal sown crop, a spacing of 20 to 22.5 cm between rows is recommended.
- ☆ If sowing is delayed, a closer spacing of 15-18 cm needs to be adopted.
- ☆ The sowing depth of wheat should be around 4-5 cm.
- ☆ There are different methods of sowing of wheat seeds such as:
    - Seed drill
    - Broadcasting method
    - Zero tillage drill
    - Rotavator

Chapter 2

AGRONOMIC PRACTICES

2.1 TYPES OF SOIL

2.1.1 Knowledge about different types of soil

Soil is one of the important resources, as fertile soil helps in producing many crops which serves the food requirements not only in the country but also in other parts of the world. Soil is not same at every place on the earth. The types of soil of a place are determined by climate, landscape, and vegetation of that place.

Different types of soil are characteristics of their color, texture and chemical properties. The soil has particles of different sizes. The relative amounts of the size of various particles in the soil determine the texture of the soil.

1. Sandy soil

- ☆ Sandy soil contains more than 60% sand and clay.
- ☆ It contains very little clay and silt, so it is porous.
- ☆ The size of soil particles in sandy soil is from 0.2mm to 2.0mm.
- ☆ The water building capacity of a sandy soil is very poor, so there appears lot of air in such type of soil.
- ☆ The soil is not good for plants.
- ☆ If water is available for irrigation then crops such as maize, millets, barley can be grown in desert soil.
- ☆ Cactus also grows in this soil.

2. Clayey Soil

- ☆ The clayey soil consists of very fine particles of clay.
- ☆ Its water holding capacity is very high.
- ☆ Wet clay soil is very sticky.
- ☆ It contains very little air.
- ☆ The size of soil particles in clay size is less than 0.2mm.
- ☆ The soil is rich in organic matter.

- ☆ The soil is not good for many plants.
- ☆ It is only good for crops like paddy, which require a lot of water.
- ☆ Clay is used for making toys, pots, and many other purposes.

3. Loamy Soil

- ☆ The loamy soil consists of sand, clay and silt.
- ☆ It contains enough hummus.
- ☆ It has a good water-holding capacity.
- ☆ It has sufficient aeration.
- ☆ It is well suited for cultivation.
- ☆ Roots of plants get enough water, air, and space to grow.
- ☆ The soil is ideal for growing crops such as wheat, sugarcane, cotton, jute, pulses, and oilseeds.
- ☆ Vegetables also grow well in this soil.

4. Red soil

- ☆ The red color of soil is due to the presence of iron oxide.
- ☆ It appears yellow when it contains less iron or more water.
- ☆ The soil contains a mixture of clay and sand, antis not fertile.
- ☆ It can be fertile by adding manures and fertilizers.
- ☆ It is suitable for growing groundnuts, pulses, millet, cotton and tobacco.

5. Black soil

- ☆ Black soil is also known as black lava soil.
- ☆ This soil black in color.
- ☆ It is formed from lava rocks and is rich in clay.
- ☆ It is ideal for growing crops such as cotton, sugarcane, tobacco, wheat, millets, and oilseeds.

2.1.3 Soil requirement in wheat

1. Wheat crop is grown in different types of soils ranging from desert soil to heavy clay soil.
2. Well drained, fertile clay loam soils having moderate water holding capacity are ideal for irrigated wheat.
3. Heavy soil with open structure which can retain the monsoon rainfall for longer period is ideal for dry wheat crop.
4. In M.S. black soils are most suitable for cultivation of wheat.

2.1.4 Soil Composition

Soil contains:

- ☆ Organic matter
- ☆ Soil organisms - Micro flora and Micro fauna
- ☆ Soil water
- ☆ Soil air
- ☆ Inorganic matter - Macro nutrients and Micro nutrients

Organic Matter

- ☆ The plants and animals grown in weathered material and the organic residues left behind decay with time and become an integral part of the soil.
- ☆ The main source of soil organic matter is plant tissue.
- ☆ Animals are subsidiary source of soil organic matter.
- ☆ The micro flora like bacteria, fungi, algae, actinomycetes, and micro fauna like protozoa, nematodes, macro fauna etc, plays an important role in formation of organic matter.
- ☆ The organic matter influences the soil in respect to colour, physical properties, supply of available nutrients and adsorptive capacity.

Soil Organisms

- ☆ Soil is the habitat for enormous number of living organisms.
- ☆ Some of the organisms are visible to naked eye while others can be seen by microscope only.
- ☆ Roots of higher plants are considered as soil macro flora while bacteria, fungi, algae and actinomycetes are considered as soil micro flora.
- ☆ Protozoa and nematodes are the significant soil micro fauna where as the earthworms, moles and ants constitutes soil macro fauna.

Soil Water

- ☆ To function as a medium for plant growth, soil should have some water.
- ☆ Promotes many physical and biological activities of soil.
- ☆ Acts as a solvent and carrier of nutrients.
- ☆ Serves as a nutrient itself.
- ☆ Acts as an agent in photosynthesis process.
- ☆ Maintains turgidity of plants.
- ☆ Acts as an agent in weathering of rocks and minerals.

Soil Air

- ☆ Oxygen is required for all biological reactions that takes place in soil.
- ☆ The gaseous phase of soil acts as a path way for intake of oxygen which is absorbed by soil micro organisms, plant roots and for escaping carbondioxide by plants.
- ☆ The two way process is called soil aeration which become important for plant growth when water content is high, as water replaces soil air.

Soil Inorganic Matter

- ☆ The inorganic constituents of soil has carbonates, soluble salts, free oxides of iron, aluminium and silica with some amorphous silicates.
- ☆ The inorganic constituents forms bulk of solid phase of the soil.
- ☆ Soils with more than 20% of the organic constituents are designated as organic soils.
- ☆ Soils where inorganic constituents are called mineral soils.
- ☆ The majority of the soils in India are mineral soils.

2.2 TESTING OF SOIL STATUS

2.2.1 Soil testing

Soil testing is a chemical process by virtue of which requirement of nutrients for plant can be analysed so as to maintain the soil fertility. It is done to analyse the fertilizer requirements for profitable wheat crop.

2.2.2 Need

Soil testing is required as:

- ☆ land which is under cropping, grain or fruit or when entire plant gets removed/harvested, in such case, the soil tends to lose a considerable amount of its nutrients. If cropping is continued over a period of time, without restoring of nutrients to the soil, its fertility will be reduced and crop yields will decline.

- ☆ nutrients also get lost from the soil by leaching and erosion, so to maintain the soil productivity at existing levels, it is required to restore the soil nutrients.
- ☆ continued maintenance of high level of soil fertility is important for profitable land use and sustained agricultural production, hence from time to time, inherent fertility of soil needs to be evaluated.
- ☆ standard soil tests normally will help in evaluating the pH, buffer pH, organic matter, cation exchange capacity, phosphorus, potassium, calcium, magnesium and base saturation.

2.2.3 Objectives

There are certain objectives of soil testing which includes:

1. promoting of organic fertilizer & insecticides on the basis of balanced use of fertilizer.
2. preparing fertility map of specific area like block or micro watershed etc based on soil testing report which will help in advocating appropriate cropping system in the required area.
3. making certain specific cropping system economically viable with respect to crop production.
4. stopping of indiscriminate use of fertilizer.
5. appropriate fertilizer dozes as recommended on basis of soil testing which includes:

- ☆ Normal soil (pH 6 to 6.5)- Nitrogenous fertilizer, Super phosphate, Ammonium phosphate etc. can be used.
- ☆ Soil with less than 6.00 pH value- Calcium amonium nitrate, Sodium nitrate, phosphoric fertilizer, super phosphate, murate of potash etc can be applied.
- ☆ Soil with more than 8.5 pH value- Urea , Ammonium sulphate, Ammonium chloride, Super phosphate , Ammonium phosphate, Murate of potash etc are used. The water logger soils can be applied Urea, Ammonium sulphate, Super phosphate , Ammonium phosphate , Ammonium chloride etc.

6. identifying quantity of organic carbon in soil as growth of crop & availability of nutrients are based on organic carbon where the ratio of organic carbon & nitrogen will be 10-12:1. The propagation of useful bacteria also depends on it.
7. recommendation of ingredient to reclaim the acidic soil or saline soils etc.
8. deficiency of micro element like Zinc, Copper, Boron, Molybdenum, Iron, Cobalt, Silicon, Manganese & Chlorine etc also adversely affect the crop condition & ultimately the production.

2.3 TESTING OF SOIL STATUS FOR VARIOUS NUTRIENTS AT AUTHORIZED SOIL TESTING CENTRES

2.3.1 Soil testing

Soil testing has been used to formulate precise recommendations for major nutrients i.e. nitrogen, phosphorus fertilization of crops in different soils and to recommend appropriate doses of amendments for salt-affected and acidic soils.

Micronutrients, comprising zinc, copper, iron manganese, boron and chlorine, though required by plants in much smaller amounts, yet are as essential for them as the major nutrients. Despite that, little attention has been paid to employ the soil testing for assessing the micronutrient status of soils and determining soils requirement for micronutrient fertilizers for growing crops. For proper determination of soil nutrients, soil should be tested at the laboratories.

2.3.2 Testing in laboratory

- ☆ Testing of soil is normally done to assess its physical properties which affect or control its behavior in relation to various purposes.
- ☆ Although testing activities include on-site/in-situ and laboratory testing, the latter one is assumed to give a better and accurate estimate of its properties.
- ☆ Each of the different properties of soil gives an estimate of a particular characteristic of it which includes moisture content of a soil specimen to guide the compaction process, besides governing the Atterberg limits, consistency, shrinkage characteristics, and a host of other strength properties.
- ☆ Size-grading of soil helps in the optimum designing for preparing concrete mixes and porosity along-with permeability tests are conducted on a soil sample to determine the suitability of semi-pervious materials for use in the core of fill dams.
- ☆ The process of soil testing requires mixing several random soil samples before submitting for testing.
- ☆ Results of the test give information to assist the homeowner in developing a fertilization program.
- ☆ Each soil testing laboratory uses different testing techniques, it is critical that the lab provide an interpretation of the soil test as a recommendation for applying or withholding fertilizers.
- ☆ The choice of plants to add to a landscape can also be directed in part through the information produced in the test.

2.3.3 Sample Selection

- ☆ The testing of soil in the laboratory requires only few grams of soil sample.
- ☆ In a homogenous field, soil samples from plough layer (0-15cm) should be selected randomly in a zig-zag manner. The samples should not be collected from near the bunds, water channels, field paths and heaps of crop straw, stubbles, manure, etc.
- ☆ The sample collected from the selected sites should be composite and mixed thoroughly in a container.
- ☆ From this lot a representative sample, about 500 gm should be taken out and air-dried under shade.
- ☆ The air-dried sample should be transferred into a clean cloth bag bearing a slip with a mention of complete address, field number, cropping sequence being followed, source of irrigation, soil type, fertilizer/manure schedule followed in the preceding crops and any other specific observation about the soil and/or the crops grown therein.
- ☆ Then the sample should be taken to the laboratory where facilities for testing soils for micronutrients are available.

2.3.4 Soil Test Report

- ☆ Many laboratories involves in standard soil test which covers set of soil chemical measurements that are important in agronomic decision making.
- ☆ Certain labs vary, typically which includes pH, buffer pH, cation exchange capacity (CEC), base saturation, organic matter (OM), and the macronutrients phosphorus (P), potassium (K^+), calcium (Ca^{2+}), and magnesium (Mg^{2+}).
- ☆ In plains and western regions where salinity and sodicity are more common, conductivity, soluble salts, sodium (Na^+) and chloride (Cl^-) may also be reported in a local laboratory's standard test.

☆ Nitrate (NO³⁻) and sulfur (S), are not usually part of standard tests, but are considered macronutrients.

☆ Nitrate tests are used in some regions for preplant nitrogen (N) management and in-season sidedress decisions, but recommendations vary from region to region.

2.3.5 Soil Test Values

pH

☆ It is a measure of acidity of a substance.

☆ It is expressed as a number, not a percentage or other quantity.

☆ A value of 7.0 on pH scale is considered neutral, the substance is neither acidic nor basic.

☆ Lower numbers on the pH scale denote increasing H^+ ions and acidity.

☆ The acidity of the soil solution affects many physical, chemical, and biological reactions necessary for plants to survive and thrive.

☆ The optimum pH value varies by crop and region, but is generally between 6.0 and 7.0 pH (slightly acidic), though alfalfa thrives between pH 6.8 to 7.0.

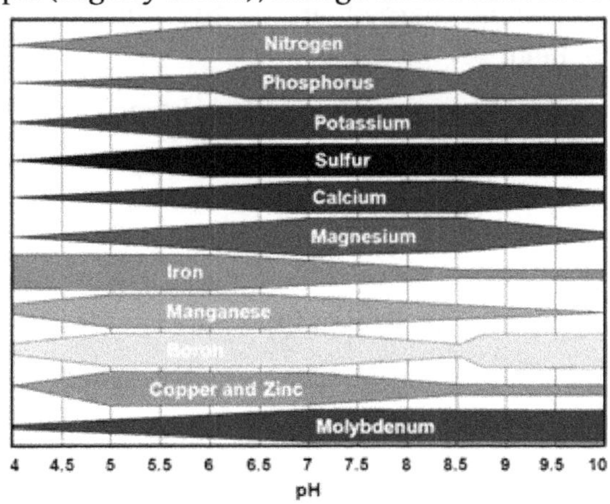

Buffer pH

- Once need for adjusting pH higher is identified, buffer pH value shows the amount of liming material required to make the adjustment.
- Low buffer pH values shows that more lime materials are necessary to raise pH than higher buffer pH values.
- Buffering refers to ability of the soil to recharge acidity, and is a function of aluminum (Al^{3+}) minerals and H^+ in the soil.
- If a soil is poorly buffered (sandy), less liming material are needed to raise pH a certain level than a soil that is highly buffered (clayey soil).
- Some soil tests may report buffer pH as **lime test index**, or **LTI**, which is merely the buffer pH value multiplied by 10.

Organic Matter

- It is expressed as a percentage and is the non-mineral content of soil sample determined by combustion.
- It has many functions, including water-holding capacity, nutrient cycling, and contributing to soil structure and CEC.

CEC

- Cation Exchange Capacity is measured in milli-equivalents per 100 grams of soil.
- It is determined primarily by soil clay mineralogy and organic matter level in the soil.
- It shows a measure of electrostatic charge sites in the soil which hold cations, or positively charged ions, like Ca^{2+}, Mg^{2+}, Mn^{2+}, Zn^{2+}, K^+, H^+, and Al^{3+}.
- For every equivalent charged site on the soil particles, a single positive charge can be held.
- In case of K^+, a K^+ ion has a single positive charge, and is held by one CEC site.
- Higher CEC values in soil represent more capacity for nutrient cations to attach.

Phosphorus P

- It is reported in **p**arts **p**er **m**illion (ppm) or pounds per acre.
- It shows an estimate of phosphorous which is available to the plants.
- The testing procedure uses extractant that is correlated with P uptake that might occur by plant roots.

- ☆ It is an important element in plants, as it is part of the DNA, RNA, the energy transfer molecule ATP, as well as amino acids and proteins.

Potassium K^+

- ☆ It is reported as ppm or lb/acre of **"exchangeable K."**
- ☆ The K^+ extraction solution measures K^+ that can be readily moved from the CEC and into soil solution, simulating the processes that might exist with a root system in the soil.
- ☆ It is vital in water regulation and enzyme activation in plants.
- ☆ In plant stover or residue on the soil surface potassium can cycle back to the soil with rainfall events and residue breakdown.
- ☆ Dry conditions also limit the movement of slowly available K^+ to the CEC.

Calcium Ca^{2+}

- ☆ Calcium content of soil is reported in ppm or lb/acre, as **exchangeable calcium.**
- ☆ It is a common element in a lot of mineral soils, and though they may occur in some regions, deficiencies are rare in most soil environments above pH 5.5.
- ☆ The easiest correction to a suspected Ca^{2+} deficiency is raising soil pH with a liming compound, and Ca^{2+} is considered sufficient at ≥200-300 ppm.
- ☆ Calcium is essential to plant cell wall structure, cell division, and many enzymatic processes.

Magnesium Mg^{2+}

- ☆ Magnesium levels are measured in ppm or lb/acre as **exchangeable magnesium.**
- ☆ **It** is a key element in chlorophyll and many enzymes and enzyme activation.
- ☆ Deficiencies can occur in crops, and can be remedied with Mg-containing fertilizers or dolomitic limestone. Mg^{2+} is sufficient for most crops at ≥50-100 ppm.

2.4 PRE SOWING IRRIGATION

A pre-sowing irrigation was applied for proper seed germination.

In this:

Zerp till-drill

Bed planting

PTOS

Farmer's practice (FP)

- ☆ A high yielding variety of seed can be sown using zero till drill, bed former, PTOS and farmer's practice at the rate of 120 kg/ha.
- ☆ Seeding is done without land preparation by zero till-drill.
- ☆ Bed can be formed and seeding can be done without prior land preparation.
- ☆ Bed was formed two rows accommodating wheat.
- ☆ Furrow to furrow was 55-56 cm.
- ☆ Bed height was 12-13 cm.
- ☆ Irrigation water was applied to furrows between beds.
- ☆ Ploughing, seeding, and leveling were done through one pass full tillage by PTOS.
- ☆ Tractor operated rotavator was used in farmer's practice for broadcast sowing.
- ☆ Data on sowing cost by tractor operated rotavator were collected from different locations.

- ☆ Seeds were treated with Provax-200 @ 3 g/kg of seed.
- ☆ Fertilizers were applied at the rate of N100+P26+K40+S20+B1 kg/ha.
- ☆ Two–thirds of N and total amount of other fertilizers were applied at sowing and remaining N was top dressed after first irrigation.
- ☆ Weeding was done at 25-27 DAS by 2-4-D amine @ 1200 ml/ha.
- ☆ At maturity, two samples of each plot in 2 m2 for zero till-drill, PTOS, farmer's practices and 2 x 1.1 m for bed planting were harvested.
- ☆ After threshed and cleaned, grain yields were recorded at 12% moisture content.
- ☆ Data on other plant characteristics during harvesting from each of the experimental plot were recorded, such as spike, plant height, spike length, spikelets/spike, grains/spike, thousand grain wt (TGW), and grain yield were recorded.
- ☆ Data were subjected to analysis of variance to sort out significant difference among treatments.
- ☆ DMRT was used to compare treatment means.

2.5 PREPARATION OF LAND FOR SOWING

2.5.1 Land levelling

- ☆ Land leveling is one of the important agronomic practice to remove ups and downs of the soil and to get better seed bed conditions for uniform distribution of irrigation water and better supply of other inputs to grow wheat crop.
- ☆ Wheat requires fertile loam to clay loam soils with good drainage for better yields.
- ☆ The wheat crop growing areas should be divided into small plots of half an acre along with channels in the middle to separate each plot where proper use of irrigation water should be there for crop needs.

2.5.2 Land Preparation

Wheat crop requires a well crushed but compact seed-bed for good and uniform seed germination. In irrigated areas, after the harvest of previous crop, the field should be ploughed with disc or mould board plough.

The purpose of land preparation is to provide the necessary soil conditions which will enhance the successful establishment of the young offshoots or the tissue culture plants received from the nursery.

Aim

The aim of preparing the land before sowing involves:

- ☆ Creation of soil structure which encourages the seedlings to emerge rapidly and uniformly and allowing young plants ready access to vital resources of nutrients, water and aeration.
- ☆ Full tillage and minimal tillage systems where minimal tillage limits soil disturbance the surface layers or just shallow slot for each crop row.
- ☆ Minimal tillage usage that can be taken into the field when the soil is wet, as perhaps after a previous rice crop, it has the prime advantage that it commonly shortens the time between successive crops/rotations which can be critical for maximise yield per year.
- ☆ Incorporating any additives like lime, compost, farmyard manure and chemicals for plant nutrition and pest control depending on location that incorporate residues remaining from previous crops.
- ☆ Controlling of weeds, pests and diseases.
- ☆ Shaping of land so that irrigation water can be applied efficiently and drained effectively and so that water logging is avoided or minimised.

Land preparation methods

Once the required area for establishing the plantation is selected by wheat grower and planning operation has been finalised, the actual preparation of land can be started which can be divided to structure and pace the implementation process in order to be ready for planting at the most suitable time, according to the specific regional climatic conditions.

1 Mechanical field preparation

The mechanical or initial soil preparation concerns mainly the preparation of a field for further detailed preparation such as irrigation system installation, hole preparation, if applicable to the area, include:

(i) debushing/bush clearing;

(ii) removal of stones and rocks;

(iii) ripping;

(iv) levelling of soil.

2 Irrigation system installation

- ☆ The type of irrigation system to be used will be determined by the availability of water, topographical and soil conditions.
- ☆ When the initial soil preparation is completed, the installation of the required irrigation system will be implemented according to the prescribed design.

2.5.2 Planking the field after each ploughing

- ☆ Planking is done to crush the hard clods to smoothen the soil surface and to compact the soil lightly.
- ☆ The field is made ready for sowing after ploughing by harrowing and planking.
- ☆ Planking should be done early in the morning.
- ☆ A buffer zone of 10-30 m width should be kept to avoid contamination from the non-organic field.

2.6 SEED RATE

2.6.1 Seed rate

- ☆ Seed rate plays an important role for optimum plant densities which is pre-requisite for increased seed yield.
- ☆ It influences the yield and yield attributes of wheat.
- ☆ High seed rate produces more plants in unit area resulting in less intra-crop competition and affecting the yield and production cost.
- ☆ Lower seed rate reduces the yield drastically.
- ☆ For medium sized grain varieties, seed rate of 100 kg/ha is recommended
- ☆ For bold sized seed variety, seed rate of 125 kg/ha is recommended
- ☆ For late sown seed sized, seed rate of 125-150 kg/ha is recommended

2.6.2 Factors

Seed rate depends on following factors:

- ☆ Soil type
- ☆ Variety

- Sowing time
- Soil moisture content at sowing time
- Sowing method

2.6.3 Considerations

- Seed should be procured from standard source.
- The seed should be cleaned thoroughly and treated before sowing.

2.6.4 Seed rate as per sowing time

- The optimum time of sowing for wheat is when the mean daily temperature is between 23±3°C and for good tillering temperature should range between 16-20°C.
- Under irrigated timely sown condition, the wheat sowing may be done in the first fortnight of November in NHZ, NWPZ & PZ and the middle of November in NEPZ and CZ.
- Under late sown conditions, the wheat sowing may be done in the first fortnight of December in NHZ, NWPZ, NEPZ and CZ whereas in the last week of November to first week of December in PZ.
- The timely sown varieties may be grown up to middle of December in NWPZ and CZ whereas in NEPZ varieties recommended for different sowing conditions may be grown for higher productivity.
- In PZ timely sown varieties may be grown across all sowing times for higher productivity.

2.7 SEED TREATMENT

2.7.1 Introduction

Seed treatments control the pathogens that occur on or in the seed. The choice of seed treatment may be dictated by whether the pathogen is borne externally or internally. These pathogens may be important for three reasons:

- Some pathogens do not survive in soil or crop residue and are dependent on the seed borne phase for survival between crops such as fungus that causes loose smut of wheat.
- If a pathogen can survive in soil or residue, being seed borne may allow it to get a head start and, thus, result in more severe disease such as fungus that causes septoria leaf blotch of wheat.
- Seed borne pathogens may hitch a ride to new localities in seed shipments.

Description of seed treatment

Loose smut:

- ☆ Treat the seed with Tebuconazole @ 1g/kg seed or Vitavax @ 2 g/kg or Bavistin @ 2.5 g/kg seed for the control of loose smut.
- ☆ Seed treatment should be done with seed treating drum.
- ☆ Seed treatment should also be done with T. viride @ 4 g / Kg seed in combination with carboxin (Vitavax 75 WP) @ 1.25 g / Kg seed or Tebuconazole (Raxil 2 DS) @ 1.0 g / Kg seed.
- ☆ Seed treatment with T. viride alone @ 4 g / Kg seed is also helpful as it reduces the rust severity.

Flag smut:

- ☆ To control flag smut, treat the seed before sowing with Thiram @ 3 g/kg or Tebuconazole 1g/kg or Vitavax @ 2 g or Bavistin @ 2.5 g/kg seed.

Other diseases:

- ☆ Treat the seed with Captan or Thiram @ 3 g/kg seed for the control of root rot, foot rot, seedling blight, black tip and black spot of glumes.
- ☆ Captan and Thiram treatment should not be done earlier than one month of sowing as it affects seed germination.
- ☆ Treat seed before sowing with Trichoderma viride @4 g/kg seed.
- ☆ It provides induced systemic resistance, thus reducing the severity of the yellow rust at later growth stages of the crop
- ☆ In termite prone areas, seed treatment with chlorpyriphos @ 0.9g a.i or Endosulfan @ 2.4g a.i/kg seed, be taken up for their management.
- ☆ Seed treatment with thiamethoxam 70WS (Cruiser 70WS) @ 0.7 g a.i./kg seed or Fipronil (Regent 5FS @ 0.3 g a.i./kg seed) is also very effective
- ☆ The IPM module involves the seed treatment with T. viride (@4g/kg seed) + carboxin (75WP) @1.25g/kg seed) or Tebuconazole (@ 1.0g/kg seed) for the control of loose smut, followed by broadcast of insecticide treated soil (with endosulfan 35 EC @ 2.3L or chlorpyriphos @ 3L/ha) at 15 DAS for termites.

Ear cockle:

- ☆ The infested seed lot should be floated in 2 percent brine solution for this purpose
- ☆ The galls will float on the surface.
- ☆ These should be separated and destroyed away from the field by burning.
- ☆ The seed should be thoroughly washed to remove the salt solution before sowing.

2.7.2 Methods of seed treatment before sowing

- ☆ Due to infection in seeds, wheat crop gets different diseases like: smut, rust, blight etc. Due to these diseases, there is adverse effect on annual production of wheat.
- ☆ The infection from such diseases could be minimized or reduced by using only healthy, and quality seeds chosen through hot water seed treatment and treatment with fungicide, which is described detail below:

(A)Hot water seed treatment

- ☆ Initially, soak the seeds in ordinary water for about 4 hours, than dip in hot water 52 ºC for 10 minutes or 55 ºC for 5 minutes.
- ☆ After that, cool and dry the seeds in shade, and let for fungicide treatment.

(B) Fungicides treatment of seed

- ☆ It is compulsory to treat the seeds for producing quality and healthy wheat crop.
- ☆ Mix the wheat seeds before sowing with Bevistin or Benlate2 grams (0.2%) per kg seeds in an air tight pot and stir it vigorously for 8-10 minutes.
- ☆ Treating the seeds this way helps in controlling diseases and germs on outer surface of wheat grain such as blast, loose smut, etc.

(c) Seed Dressing

- ☆ The most common method of seed treatment.
- ☆ The seed is either dressed with a dry formulation or wet treated with a slurry or liquid formulation.
- ☆ Dressings are applied both on-farm or in specialised seed treatment facilities.

(D) Seed Coating

- ☆ A special binder is used with a formulation to enhance adherence to the seed and begin to impact seed size and shape.
- ☆ Coatings require advanced treatment application technology.

(E) Seed Pelleting

- ☆ The most sophisticated seed treatment technology, resulting in changing the physical shape of a seed to enhance plantability and handling.
- ☆ Pelleting requires specialised application machinery and techniques and is the most expensive of the applications.

2.7.3 Procedure and precautions for seed treatment for control of termite/seed and soil borne diseases

Precautions

The following things need to be considered while treating the seeds:

- ☆ Select non-windy place to treat seeds with fungicide so that air will not carry away the fungicide.
- ☆ Children, aged people, pregnant women are more vulnerable to poisonous effect of fungicide hence; they should not be involved in treating the seeds.
- ☆ Use mask, gloves, and goggles while treating seeds.
- ☆ Wear the clothes that cover whole body.
- ☆ Use only proper or prescribed amount of fungicides.
- ☆ Wash thoroughly with water when fungicide touches your skin directly.

Protection of Seeds and Seedlings

- ☆ Seeds and seedlings are vulnerable to many soil borne and foliar pests.
- ☆ Insects and pathogens can destroy germinating seeds and young plants, which are relatively tender and lack food reserves to recover from injuries or to survive extended periods of stress.
- ☆ Under stressful conditions, a number of aggressive or even fairly weak pathogens can become active and cause plant population and yield losses.
- ☆ Seed treatments can protect the seed and seedling from attack by certain insects and pathogens. No systemic fungicides or insecticides form a chemical barrier over the surface of the germinating seed which protect the germinating seed from chewing insects, such as wireworms, or soil borne pathogens, such as pythium.
- ☆ Certain systemic seed treatments can protect aboveground parts from sucking insects, such as aphids, or foliage diseases, such as rust.
- ☆ Systemic fungicides and biological seed treatments can also protect young plants from root rot. Although the duration of protection may be limited, a delay in infection can reduce the losses. For chronic diseases, such as root rots, the earlier that the infection takes place, the greater
- ☆ will be the damage.
- ☆ Seed treatments lasts only for 10 to 14 days beyond planting, with pesticide breakdown being most rapid under warm, moist conditions.

- ☆ Certain active ingredients can protect seedlings considerably longer when applied at the highest labelled rate

2.7.4 Advantages

- ☆ Seed borne pathogens are vulnerable: The seed borne phase is weakest link in life cycle for many plant pathogens. With seed treatments to control seed borne pathogens is very effective for disease control.
- ☆ Precision targeting: Seed treatments are not subject to spray drift as chemicals are applied directly to seeds, little is wasted on no target sites like bare soil.
- ☆ Optimum timing: Seeds and seedlings are vulnerable to diseases and insects as compared to mature plants. Applying treatments to seeds will allow pesticides to be present when needed most.
- ☆ Low dose: Relatively small amounts of pesticides are used in seed treatments compared to broadcast sprays as it reduces the cost and the potential environmental impact and the probability of chemical residues in harvested grain.
- ☆ Easy to apply: Seed treatments are relatively easy and cheap to apply as compared to broadcast sprays.

2.7.5 Disadvantages

- ☆ Accidental poisoning: Treated seed looks like food to some animals as hungry livestock birds, such as pheasants or quail consume spilled treated seed.
- ☆ Cropping restrictions: Just like other pesticides, some seed treatments may have significant grazing or rotation crop restrictions.
- ☆ Limited dose capacity: The amount of pesticide that can be applied is limited by how much will actually stick to the seed. Seed coating technologies are helping to overcome this limitation, but phytotoxicity may still be a problem.
- ☆ Limited duration of protection: The duration of protection is short due to relative small amount of chemical applied to seed, dilution of chemical as plant grows, and breakdown of chemical.
- ☆ Limited shelf life of treated seed: Producing excess treated seed is undesirable as shelf life of treated seed is limited. Surplus treated seed cannot be sold for grain which is a serious limitation for seeds where seed germination and vigor decline relatively quickly.
- ☆ Phytotoxicity: Pesticide injury to plant tissues is called phytotoxicity as seed treatments exist in high concentrations on tender tissues of germinating seeds and seedlings having very low phytotoxicity.

- ☆ Worker exposure: In course of treating and handling large volumes of seed, workers may be exposed to seed treatment chemicals as aerosols. Inhalation of aerosols and skin contact with seed treatments must be prevented in the seed treatment process

2.8 SOWING TIME AND METHODS OF SOWING

2.8.1 Sowing time

- ☆ Seed is sown by drilling or broadcasting.
- ☆ Sowing of seed with seed drills, whether tractor drawn or bullock-drawn will deposit the seed at uniform depth, giving a uniform stand and lead to the early emergence of vigorous seedlings.
- ☆ Under irrigated conditions, first fortnight of November is considered as optimum time for sowing the medium long-duration wheats, e.g. 'kalyansona' all over India.
- ☆ For short-duration varieties, e.g. 'Sonalika', second fortnight of November is optimum time.
- ☆ In eastern India, the sowing of wheat is delayed due to late harvesting of paddy which can go up to even the third week of December.
- ☆ In north- western India, wheat sowings get delayed due to late harvesting of paddy or sugarcane or potato.
- ☆ Wheat sowings beyond the rain fed conditions, second half of October is good time.
- ☆ Sowing of wheat, when the temperatures is high, results in poor stand, poor tillering, attack by root-infecting fungi and pests and the early onset of flowering.
- ☆ The success of rained wheat depends on the amount of conserved moisture in the soil.
- ☆ In general, the maximum productivity potential of promising wheat varieties under irrigated conditions is attained by sowing them between 10th and 25th November in all the climatic zones other than the peninsular zone.
- ☆ Rainfed wheat is generally sown during the second fortnight of October to early November.
- ☆ Under specific circumstances, wheat is sown in the month of December too.
- ☆ In late sown wheat, only short duration varieties should be used.

2.8.2 Factors on which Sowing time depends

The optimum sowing time of wheat varies over the growing areas which depends upon

- ☆ type of variety,
- ☆ weather conditions,
- ☆ soil temperature,
- ☆ irrigation facilities
- ☆ land preparation.

2.8.3 Types of wheat with sowing time

Dwarf wheat:

- ☆ Normal sowing time: Last week of October to 3rd week of November
- ☆ Late sowing time: Up to last week of December

Tall wheat:

- ☆ Normal sowing time: Last week of October to first week of November.
- ☆ Late sowing time: 2nd week of November

Durum wheat:

- ☆ Last week of October to first week of November

Important:

- ☆ If possible wheat must be sown at right time as it will give maximum yield.
- ☆ A delay of one week in wheat sowing reduces the yield by about 150 Kg/acre.
- ☆ The varieties recommended for late sown condition must be sown if the wheat sowing has been delayed beyond November.
- ☆ Dwarf wheat should be sown when average temperature is 22^0

2.8.4 Different methods of sowing

Planting method plays an important role in the placement of seed at proper depth, which ultimately affects crop growth. The selection of suitable sowing method plays an important role in the placement of seed at proper depth, which ensures better emergence and subsequent crop growth. Wheat is planted with different sowing methods depending upon the available soil water, time of planting, amount of residue in the field and availability of planting machine.

Broadcasting:

- ☆ The seeds are broadcast and mixed in the soil using harrows.
- ☆ The uniform seed distribution cannot be achieved leading to irregular plant stand and poor germination.
- ☆ If broadcasted seeds are not covered properly, they are picked up by birds resulting in wastage of seeds.

Behind Local Plough

- ☆ The seeds are dibbled by hand into the furrows that are opened with the plough known as kera (hand) method.
- ☆ On dropping through Pora or Nai or Hazara, a special attachment with local plough is called Pora method.
- ☆ The seeds are dropped at a depth of 5 to 6 cm.
- ☆ The germination is satisfactory under this method of sowing.

Drilling:

- ☆ Before sowing seeds by drilling, seed-bed should be thoroughly prepared.
- ☆ The seed bed should be well leveled and free from clods and weeds.
- ☆ The seeds are sown by seed drilling at uniform depth.
- ☆ It results in uniform germination and regular stand.

Dibbling:

- ☆ In case of limited seed availability, dibbling method is used for sowing.
- ☆ Dibblers have a wooden or iron frame with pegs.
- ☆ The frame is pressed in the field and lifted and then one or two seeds are dropped by hand in each of the hole.
- ☆ It is a time consuming process.

Zero Tillage Technique:

- ☆ It is a useful method under rice-wheat cropping system.
- ☆ Once the rice crop is harvested, field requires at least 6-8 tillage operations to make field ready for sowing of wheat.
- ☆ A zero-till-seed drill machine is used for direct sowing of wheat in rice field without field preparation which helps in advancing of sowing of wheat.
- ☆ Under zero tillage, there should be proper moisture in the field at the time of sowing.

- ☆ The rice crop should be harvested close to the ground.
- ☆ The field should be free from weeds.
- ☆ Sowing depth of about 5 to 6 cm should be maintained.
- ☆ Light plank may be used behind the machine.

2.8.5 Care to be taken while sowing wheat

Whatever the soil moisture conditions at wheat planting time, there are a few important steps which a wheat cultivator should take in order to improve their chances of getting a good sowing of wheat which includes:

- ☆ Proper tractor speed: It is best to use a tractor speed of between 5 and 6 miles per hour in most cases when drilling wheat as per the amount of down pressure on the openers. If higher speeds are used, the openers can tend to "ride up" in the soil every now and then if down pressure is insufficient.
- ☆ Proper, uniform seeding depth: The ideal planting depth for wheat in many cases is about 1.5 inches. When planting early into very warm soils, it is especially important not to plant too deeply as coleoptiles lengths are shorter than normal under warm conditions. On the other extreme, producers should also be especially careful not to plant too deeply when planting later than the recommended time into very cool soils. Getting a uniform seeding depth is also important. Where wheat cultivator is planting into fields with heavy residue, or where there is uneven distribution of chaff from the previous crop, uneven planting depth can be a serious problem.
- ☆ Firm seedbed: Planting into loose, fluffy soils can be a problem where soils have been tilled repeatedly during the summer. When seeds are planted into loose soils, rains in the fall will settle the soil and leave the crowns of the seedlings too close to the soil surface.
- ☆ Optimal planting time: In general, wheat should be planted somewhere around Hessian fly-free date. There may be good reasons to plant some wheat before the fly-free date, such as planting for pasture or time pressures from having considerable acreage to plant. When planting late, seeding rates should be increased by 25% to 50% to help ensure an adequate stand and compensate for the lack of tillering.
- ☆ Adequate soil fertility: Wheat cultivator should apply at least part of their nitrogen before or at planting time to get the plants off to a strong start. Nitrogen rates of 20 to 30 pounds can help with fall establishment and tillering. If the soil is low or very low in phosphorus or potassium, these nutrients should be applied at planting time as well so that the plants benefit

early in their development. Starter phosphorus with the seed or band-applied close to the seed can also help with fall early growth and establishment, particularly in low-testing soils.

- Row-crop stubble planting adjustments: When planting wheat into grain sorghum stubble, producers will need an extra 30 pounds of N per acre over their normal N rate. Also, it is important to make sure the sorghum is dead before planting wheat. When planting wheat into soybean stubble, producers should not reduce their N rates since the N credit from soybeans doesn't take effect until the following spring. If the wheat is being planted no-till after row-crop harvest, N rates should be increased by 20 pounds of N per acre over the normal N rate. Seeding rates should be increased when planting wheat late after row-crop harvest.
- Potential disease issues: Wheat cultivator should watch out for potential disease issues when planting into corn residue. The risk of some diseases may be higher when wheat is planted into fields with large amounts of corn residue left on the soil surface.
- Use of a seed treatment

2.8.6 Avoid wheat sowing by broadcasting

Sowing of wheat through broadcast method should be avoided due to following reasons:

- In this method, seeds are broadcast and then worked in by harrowing in order to cover it.
- In this, seeds are not uniformly distributed in the field, so such method of sowing is very insufficient and should not be encouraged.
- Germination of broadcast seeds is relatively poor and the plant stand is often irregular.
- Wastage of seed also results as most of the seed is left on the surface where they cannot germinate and may therefore, be picked up and eaten by birds.
- The method may not ensure that all seeds are sown at the correct depth.
- Incorrect depth, if too deep, would result in germination that would not allow the young plant to break the surface of the soil and prevent sprouting.
- If the seeds are not sown evenly then there would be a lack of various nutrients from sunlight, oxygen etc in many crops or plants.

2.8.7 Calibrate the machine properly before sowing.

- Calibration of equipment by trial and error over several acres into planting can be costly in many ways and quite profitable for your seed supplier.
- Seed lots and species vary in their flowability.

Calibration of seeding equipment:

Before taking the seed drill or crop planter in the field, wheat cultivator should calibrate the metering mechanism for obtaining required seed rate per hectare or per acre of the crops to be planted.

The following steps are followed in calibration of seed drill:

1. Put up the seed drill and mark a reference point on the ground wheel.
2. Measure the diameter of ground wheel.
3. Measure the working width of the drill or planter.
4. Workout the circumference of ground wheel and multiply it with diameter.
5. Assume the planter or drill used in a field with certain length and breadth.
6. Workout the revolution of the ground wheel required to travel certain distance.
7. Collect the seeds from all the furrow openers separately and weigh all the collected seed separately.
8. See that there should not be large variation in the weight of seed collected from different furrow openers.
9. Find the total no. of revolutions required to cover one hectare of field using the formula, $(100 \times 100)/(\pi x D x W)$
10. Allow 10% slippage of ground wheel under actual field conditions and again use the formula
11. Find the total amount of seed for above revolutions and using values of revolutions, obtain the quantity of seed to be sown in the field.
12. Repeat the process by suitably adjusting the lever on indexing device till a desired seed rate recommended for particular crop under planting or sowing is achieved.

2.8.8 Zero till machine preferred in paddy-wheat cropping system

Zero tillage

- ☆ It is the practice of not adopting any tillage operations or minimum tillage operations in soil so that it is least disturbed.
- ☆ In this technique there is no preparation of soil prior to sowing or practice in which soil disturbance is limited only to the spot where seed would be placed.

Objectives:

- ☆ Management of crop residues
- ☆ Increasing or decreasing infiltration

- ☆ Killing weeds
- ☆ Breaking soil crusts
- ☆ Breaking root restricting layers
- ☆ Incorporating/placing fertilizers and amendments
- ☆ Modifying thermal regime of soil
- ☆ Reducing the insect and disease incidence
- ☆ Minimizing erosion losses by wind and water

Zero Tillage for Cropping Systems

- ☆ Through various experiments it has been concluded that use of zero tillage in certain cropping system gives increment.
- ☆ Wheat sown by zero tillage after harvest of rice crop.
- ☆ Wheat taken after soybean/maize/sorghum.
- ☆ Pulse crop can be planted with zero till without any problem.
- ☆ Green gram can be planted by this technology immediately after the harvest of wheat which can be used for grain and green manure as well.

Zero Tillage Machine

- ☆ The Zero Tillage machine is good to sow crop immediately after the rice harvesting without going in for field preparation.
- ☆ The knife type tines in the machine are used for cutting open the slits in the soil to place seed at appropriate depth.
- ☆ The machine has balancing wheel on one side and a rigid driving and balancing wheel on the other side.
- ☆ If the surface of field after harvesting of Rice becomes uneven, the rigid driving wheel of the machine skips cutting the soil and place the seed at places of depressions which results in poor crop strength.
- ☆ The machine improves further to overcome problem of patchy of crop stand by using two balancing wheels on sides and driving wheel in front of the drill.
- ☆ The driving is providing with spring action to take care about uneven field.
- ☆ The upgraded machine shows uniform crop stand and serves as improvement over the earlier Zero tillage machine.

Chapter 3

INTEGRATED NUTRIENTS/ FERTILIZER MANAGEMENT

3.1 SOIL SAMPLING PROCEDURE

3.1.1 Introduction

Soil sampling is done for checking of soil pH, phosphorus, potassium and micronutrients which should be done after every three to four years as these values do not change rapidly even with fertilization. Soil samples for soil nitrate should be taken 1-3 months before planting each wheat crop. In most of dryland or rainfed wheat production areas, wheat root will have depth of 5-foot with full profile of soil water.

Effective and profitable fertilizer recommendations require accurate sample of the field. It is very important that the soil sample should accurately represent the field to be fertilized. Previous crop history, previous lime and fertilizer rates and soil characteristics will cause variance in results if not accounted while sampling.

- ☆ Soil sampling is most challengeable work, as few grams of soil sample represents for a given area, hence soil samples needs to be taken in such a manner that the collected sample should reflect the true fertility of soil for any part of the land area.

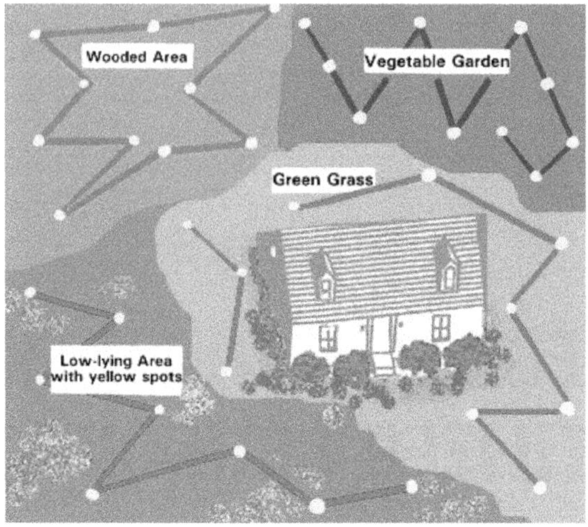

- ☆ If the field is levelled and soil appears to be uniform, only one composite sample if taken properly could be enough for an area of 4-5 ha.
- ☆ Variation in slope, colour, texture, crop growth, unusual spots and management practice should be taken into account and separate sets of composite samples must be collected from each such area.
- ☆ Recently fertilized plots, bunds, channels, marshy tracts and area near trees, wells, cattle dung and compost piles or other non-representative locations must be avoided during the sampling.

3.1.2 Sampling Tools:

There are many tools that are used for soil sampling which includes:

- ☆ spade/khurpi/proble

- ☆ auger (tube and screw type),

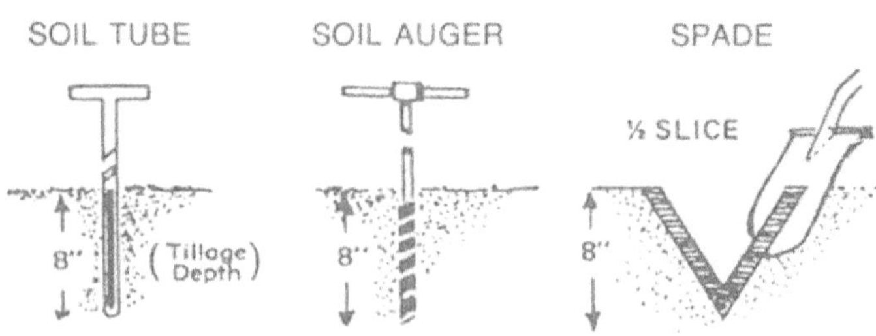

- ☆ plastic bucket,

- ☆ plastic bag,
- ☆ scale
- ☆ water proof marker

In case of soil sampling is done for soft and moist soil, the spade or khurpi is quite satisfactory while screw type auger is used for hard/dry land.

Soil probe/spade is the best tool for obtaining soil samples. For getting soil sample:

- ☆ Push the probe/spade to 8 inch depth and remove the core.
- ☆ Take a slice of soil about 1/2 inch thick, with the spade.
- ☆ Using a trowel, cut a 1/2 core lengthwise from the center of the slice while it is still on the spade. Put the core in a bucket to be mixed with the 15 other cores from the designated sampling area.

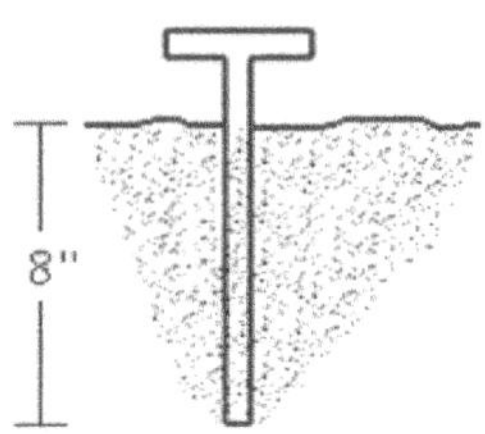

3.1.3 Selection of soil sampling area

- ☆ Sampling of soil is done by visual survey of the field where variation related to slope, colour, texture, management and cropping pattern are observed.
- ☆ The land selected should be demarcated into uniform portions so that each demarcated portion should be sampled separately.

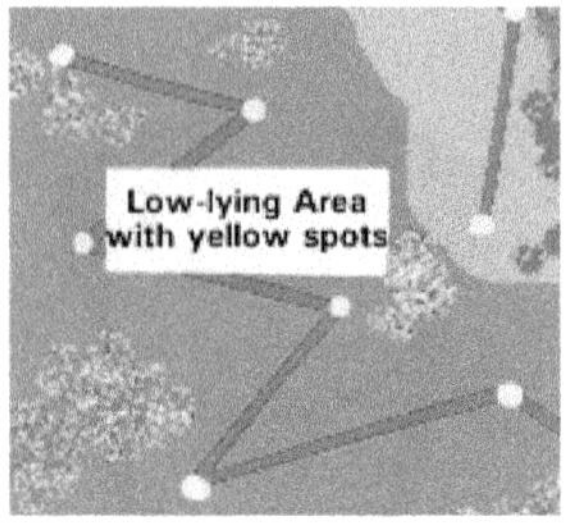

- ☆ If all conditions are same, then one field is treated as single sampling unit.
- ☆ The land selected should not be more than 1 to 2 hectares, and should be an area where a farmer is willing to give extra attention.

3.1.4 Sampling Depth

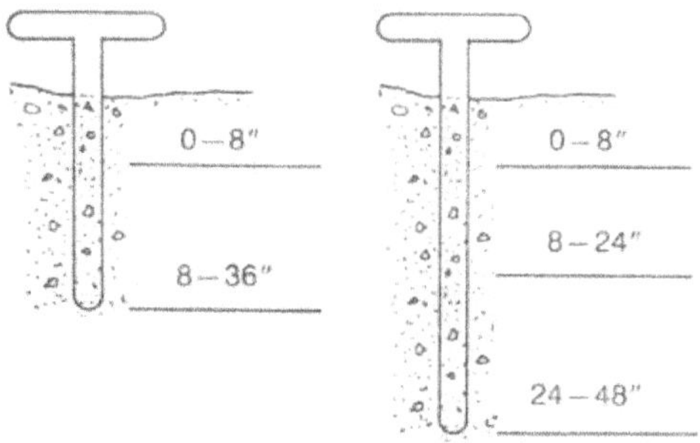

- ☆ Surface soil samples from fields should be taken to a depth of 8".
- ☆ Subsoil samples should be taken from 8" to 36" for residual nitrate.
- ☆ For better sampling, take sample from 8-24" and 24-48".

3.1.5 Sampling procedure

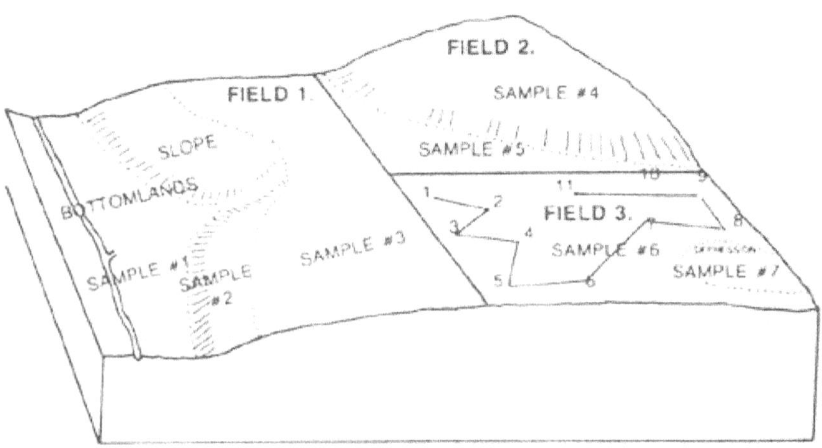

- ☆ If fields area is large, then select the part of field and divide into respective field area. If soil areas in field is different in appearance (slope, drainage, color or texture), then each soil area should be sampled separately.
- ☆ Clear the area for sampling by removing or scrapping away stones, rubbish or trash from surface to expose the soil before sampling.
- ☆ Once the land is demarcated, mark the portion of land in zig-zag manner by different sampling unit boundaries.

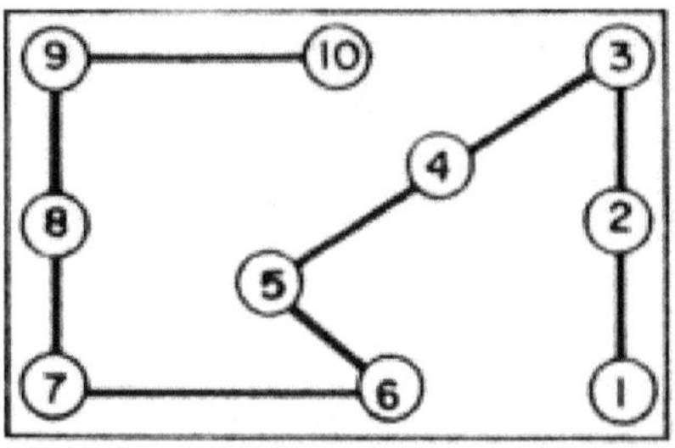

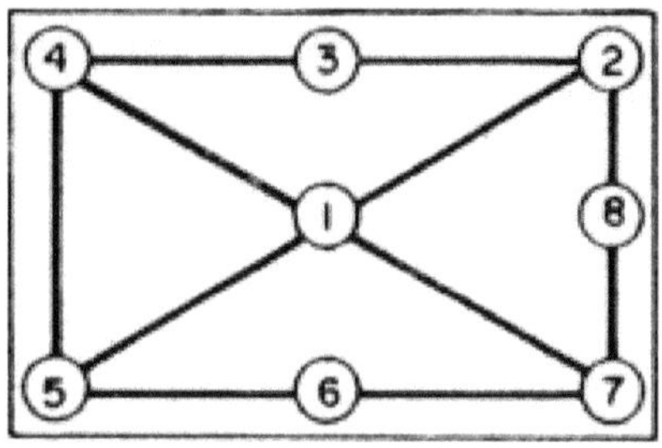

- ☆ On the plain with number of samples and manner of composite, enter the sampling and mark each unit with different letters/numbers.
- ☆ Now take a slice of the plough layer cut at intervals of 15 to 20 steps or as per area to be covered.
- ☆ Now remove away surface liter and mark the sample area from the soil sample to be taken.

- ☆ Using spade or trowel, get the soil sample at specified depth, i e 2.5 cm thick and 5 cm wide.

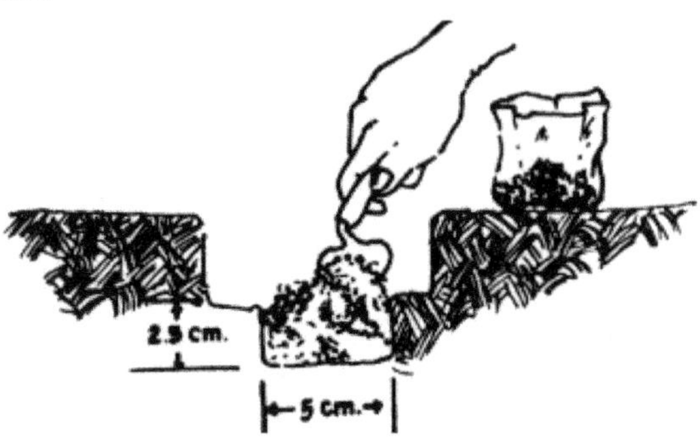

☆ Mark V shaped cut to a depth of 15 cm in the sampling spot with the help of spade to remove 1 to 2 cm slice of soil.

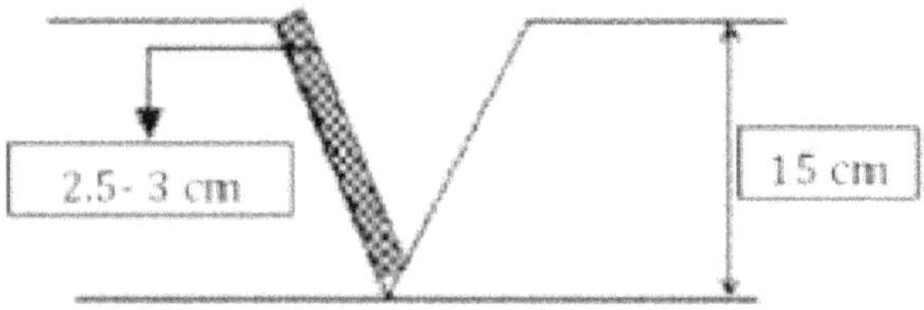

☆ Now collect the sample on the blade of the spade and put in a clean bucket.

☆ Collect soil sample and place it in a clean bucket or tray.

☆ Now take samples from 10 to 20 spots as marked for one composite sample as per size of the field.

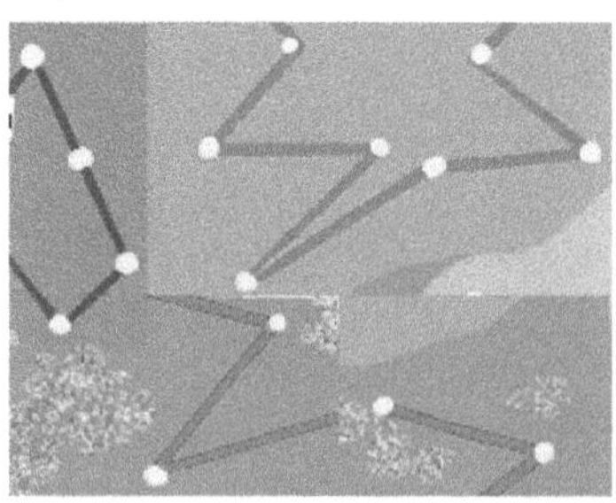

☆ Collect the different soil samples with the help of augur from the plough depth and keep them in separate glass jar.

- ☆ The glass jar should be marked with any reference, so that the samples can be easily identified.

3.2 PREPARATION OF SOIL SAMPLES

Preparing the collected soil sample for testing in laboratory involves the following steps:

- ☆ Soil samples which are collected from different units usually contain moist, aggregated state, etc that is not suitable for laboratory analysis.
- ☆ The soil sample is collected in jar/bucket should be free from organic debris, rocks and trash and needs to be mixed properly.

- ☆ Pour the soil sample from the bucket/jar on a piece of clean paper or cloth and mix it thoroughly.

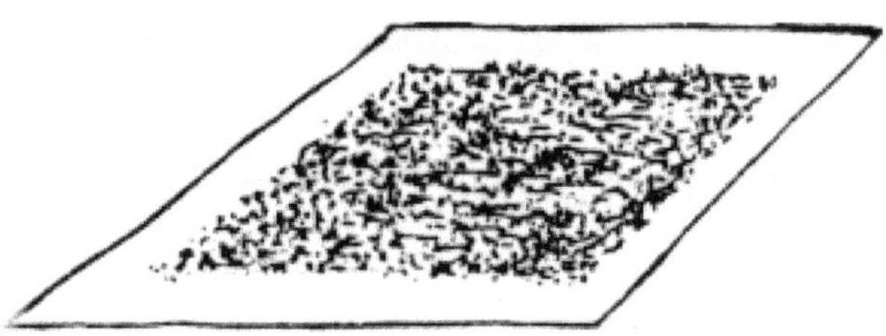

- ☆ The soil sample spread on a plastic or on thick paper should be dried and then grinded to avoid contamination.

☆ Field moist samples before drying, should pass through a 6 mm sieve by rubbing with fingers.

☆ After sieving, samples are air-dried (25-35° C; relative humidity 20-60%) and stored.

☆ Samples should not be exposed directly with the sunlight, but should be kept under the shade for drying.

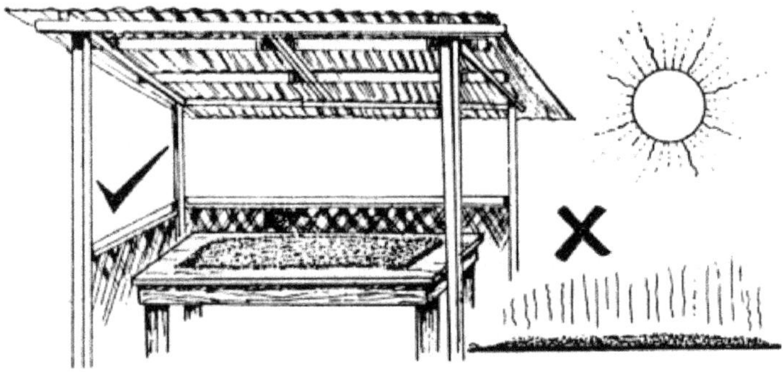

☆ To speed up the drying process, break the samples into different compartments.

☆ Keep the dried soil sample in a clean plastic bag free from any contamination.

☆ Write details of the sample in the information sheet and put it in the plastic bag.

☆ Tie the mouth of the bag carefully.

3.3 SUBMISSION OF SOIL SAMPLES FOR ANALYSIS

3.3.1 Soil Testing laboratories

Soil test determines soil nutrient capacity by mixing soil during the analysis with strong extracting solution. The soil reacts with the extracting solution, thereby releasing some nutrients. As soil supplies most of the mineral nutrition for higher plants through plant's root system, the extracted nutrient concentration is evaluated as per research that relates plant utilization to soil nutrient concentrations.

Soil testing laboratory is mainly used for testing soil samples which provides an extension advisory to the farmers by giving soil testing reports with advice and recommendations on soil condition and suitable crop that can be cultivated on particular soil.

Soil sample collected from different field units are supplied to soil testing laboratories. Standard or routine soil tests vary from laboratory to laboratory which includes analysis about:

- ☆ soil texture;
- ☆ electrical conductivity
- ☆ soil pH;
- ☆ available phosphorus (P),
- ☆ potassium (K),
- ☆ calcium (Ca),
- ☆ magnesium (Mg);
- ☆ sodium (Na);
- ☆ cation exchange capacity
- ☆ OM content

As part of soil analysis, laboratory usually supplies some interpretation, which covers indication of whether individual soil tests are low, medium or high. The laboratory provides fertilizer recommendations based on the analysis, as these recommendations are plant and soil specific.

Soil analyses are relates to potential nutrient uptake, supplementation of plant nutrients through fertilisation and target yield.

Soil testing laboratory serves as production line raw materials where soil samples are received at one end and finished products report emerges at other end. The production line has series of operation that should be done in logical sequence with operations simultaneously.

3.3.2 Facilities and Equipments in Laboratories

Soil sampling laboratories have following physical facilities:

a. Soil preparation room

b. Sample storage space

c. Main laboratory area

d. Research room

e. Chemical storage room

f. General office and hall

Soil sampling laboratories have following equipments and instruments:

1. Conductivity bridge
2. Shaking apparatus
3. PH meter
4. Rincing and stirring
5. Photoelectric colorimeter
6. Flame photometer
7. Centrifuge
8. Analytical balance.
9. Triple beam balance
10. Multiple dispensing equipment
11. Washing assembly
12. Pipetting assembly
13. Calculating machine
14. Constant holtage stabilizer
15. Water distillery
16. Oven
17. Air compressor

Materials and Equipments required in soil sampling laboratories include:

a. samples sheets with result.

b. Electrodes for PH meter.

c. Glass atomizer for flame photometer.

d. Conductivity Cell

e. Lamps – for colourimeter

f. Filter for colourimeter

g. Standard solutions for phosphorous, potassium and organic carbon.

h. Glasswares

i. Chemicals

j. Weighing scoop

k. Campe hairbrushes

l. PH meter

m. Colourimeter

n. Flame photometer

o. Conductivity bridge

p. Vacuum pump

q. Platining kit

r. Shaking machine

3.4 COLLECTION OF SOIL TESTING REPORTS

3.4.1 Soil test reports

- ☆ Soil test reports from the lab provide a representation of soil sample results in which information about different levels of various nutrients are described.
- ☆ In this, optimal and excessive levels mean that the nutrient concentration in the soil is more than adequate for optimum crop growth.
- ☆ The data obtained from soil analysis would be meaningless till they are correlated with crop yield.
- ☆ On the basis of soil testing results, the recommended fertilizer does for the crops are modified for making fertilizer recommendations to the farmers.

3.4.2 Format:

Once the soil test report gets prepared, it is the duty of wheat cultivator to collect the report and understand the basic information for cultivating wheat crop. In order to understand the report, the wheat cultivator should study each and every column of the report.

SOIL ANALYSIS REPORT

E INFORMATION SHEET: 123456

LAB NUMBER	SAMPLE IDENTIFICATION	ORGANIC MATTER		PHOSPHORUS						POTASSIUM		MAGNESIUM		CALCIUM		SODIUM		pH		CATION EXCHANGE CAPACITY	PERCENT BASE SATURATION (COMPUTED)				
		MODIFIED LOI		P_1 (WEAK BRAY)		P_2 (STRONG BRAY)		BICARBONATE P		K		Mg		Ca		Na		SOIL pH 1:1	BUFFER INDEX	C.E.C.	% K	% Mg	% Ca	% H	% Na
		percent	RATE	ppm	RATE	ppm	RATE	ppm	RATE	ppm	RATE	ppm	RATE	ppm	RATE	ppm	RATE			meq/100g					
F LAB NUMBER	G SAMPLE IDENTIFICATION	(1)				(2)				(3)			(4)			(5)		(6)	(7)	(8)			(9)		

LAB NUMBER	NITRATE-N										SULFUR		ZINC		MANGANESE		IRON		COPPER		BORON		EXCESS LIME RATE	SOLUBLE SALTS		
	SURFACE			SUBSOIL 1			SUBSOIL 2			Total lbs/A	S ICAP		Zn DTPA		Mn DTPA		Fe DTPA		Cu DTPA		B SORB. DTPA			1:1		
	ppm	lbs/A	depth (in)	ppm	lbs/A	depth (in)	ppm	lbs/A	depth (in)		ppm	RATE	ppm	RATE	ppm	RATE	ppm	RATE	ppm	RATE	ppm	RATE		mmhos/cm	RATE	
			(10)								(11)		(12)		(13)		(14)		(15)		(16)		(17)	(18)		

The format of soil test report includes:

Soil Analysis

1. Organic Matter

Percent organic matter is a measurement of the amount of plant and animal residue in the soil. The color of the soil is usually closely related to its organic matter content, with darker soils being higher in organic matter

2. Phosphorus

Shows three types of phosphorus tests; P1 (weak Bray), P2 (strong Bray) and Bicarbonate P (sodium bicarbonate).

3. Potassium

It measures the available potassium in the sample where its optimum level varies with crop, yield, soil type, soil physical condition, and other soil related factors.

4. Magnesium and Calcium

It shows the levels of calcium and magnesium in the soil that are affected mainly by soil type, drainage, liming and cropping practices.

5. Sodium

It shows the sodium level in soil sample

6. Soil pH

This shows soil pH measures, active soil acidity or alkalinity in soil sample.

7. Buffer Index

It shows an index value for determining the amount of lime to apply on acid soils with pH less than 7.0.

8. Cation Exchange Capacity

It measures the soil's ability to hold nutrients like potassium, magnesium and calcium as well as other positively charged ions as sodium and hydrogen in soil sample.

9. Percent Base Saturation

It shows part of Cation Exchange Capacity occupied by given cation in soil sample.

10. Nitrate-Nitrogen

The soil test measures nitrate-nitrogen (NO3-N) which is water soluble and readily available for plant uptake.

11. Sulfur

The soil test measures everal forms of sulfur that can be readily available.

12. Zinc

It shows extraction of the zinc in soil sample.

13. Manganese

It shows detail about manganese extraction reading using DTPA extraction process.

14. Iron

It shows details about iron extraction using the DTPA extraction process in the sample.

15. Copper

It shows detail about copper extraction using DTPA extraction process in the sample.

16. Boron

It shows reading of boron extraction from the soil using DTPA/Sorbitol where adequate levels shows from 1 to 1.5 ppm.

17. Excess Lime Rate

It shows detail about visual rating of free lime where soil with high amounts of free lime have problems with availability of major and minor elements to the plant.

Office Information

A. Report Number

All samples are filed by report number. When contacting our lab concerning a certain report, be sure to refer to this number.

B. Account

An account number has been assigned to each client. The use of this number will speed up the processing and location of samples within the laboratory system.

C. Report Date

The date which the sample data was reported is shown here.

D. Received Date

The date which the sample was received at the laboratory appears here.

E. Information Sheet Number

The number of the information sheet which was submitted with the samples in this report is listed here.

F. Lab Number

The identification number which was assigned by the laboratory to each individual soil sample is shown here. There may be more than one laboratory number per report.

G. Sample Identification

The identification number assigned by the client to each individual sample is reported here. Because of limited space, samples numbers must be limited to 10 digits.

3.5 APPLICATION OF ORGANIC AND INORGANIC FERTILIZERS

3.5.1 Organic fertilizers

- ☆ It comprises of variety of plant-derived materials that ranges from fresh or dried plant material to animal manures and litters to agricultural by-products.
- ☆ The nutrient content of organic fertilizers varies greatly among source materials, and readily biodegradable materials make better nutrient sources.
- ☆ Nitrogen and phosphorus content is lower, often substantially lower, in organic fertilizers compared to chemical fertilizers.
- ☆ Moisture content is another factor that reduces or dilutes the nitrogen and phosphorus concentrations of organic fertilizers.
- ☆ It provides energy for biological activity.
- ☆ Many of the nutrients used by plants are held in organic matter until the organisms decompose the materials and release them for the plants' use.
- ☆ It attracts and holds plant nutrients in an available state, reducing the amount of nutrients lost through leaching.
- ☆ It improves soil structure, so that air reaches plant roots and the soil retains moisture.
- ☆ The organic matter and the organisms that feed on it are central to the nutrient cycle.

3.5.2 Determine the right organic and inorganic fertilizers to be applied

- ☆ Inorganic fertilizers in combination with organic fertilizers help in improving soil health and sustain crop yield.
- ☆ For better crop production, both organic and inorganic plays an important role.
- ☆ Fertilizer use, though has been a prime contributing factor for the increased agricultural productivity in the areas, there are situations when they are not available in right time and adequate quantity despite high priced.
- ☆ After the soil test results, wheat cultivator can be able to judge the required supplemental fertilizer along with type of fertilizer to be used.

The factors which involves in selecting correct fertilizer include:

- ☆ physical properties
- ☆ chemical properties
- ☆ environmental stewardship
- ☆ economics

The good fertilizer selection involves:

- ☆ Solids, liquids or gases forms where each physical form has its own uses and limitations that provides the basis for selecting the best fertilizer for the field.
- ☆ Correct solubility, as it shows how readily the nutrients are dissolved in soil water and are taken by the plants.
- ☆ Required salt index as soluble salts at high concentrations in soil solution will injure/kill plants or prevent germination of seeds.
- ☆ Checking of acidity of some fertilizer materials.
- ☆ Correct particle size as it is important for producing stable, high-quality blended fertilizer.
- ☆ Comparison of material price per actual plant food by formula, (Price per ton of fertilizer)/(2000 x plant food content as decimal value) = per rupee of plant food

3.6 PREPARING FIELDS APPROPRIATELY FOR THE FERTILIZER APPLICATION

Fields should be prepared for fertilizing as it improves the growth of crop. For preparing the land follow the steps:

- ☆ Initially do the soil testing of the land to determine soil's level of nitrogen, phosphorus, and potassium and soil's pH.
- ☆ Remove large stones to make the field easier for tillage.
- ☆ Loose the soil by pitch fork or shovel by digging 12–14 inches in ground and then turning the soil over for loosening.
- ☆ Use recommended organic material over the surface of the soil and with tiller or pitchfork, evenly distribute the area.
- ☆ Using tiller or pitchfork, mix the compost with the soil so that the soil can benefit from the added nutrients in the compost.

- ☆ Use tiller lightly across the surface of the soil and organic material to level out the soil which makes easier for planting.
- ☆ Now apply fertilizer for planting of crop.

3.7 MIX THE FERTILIZERS IN APPROPRIATE QUANTITIES

- ☆ A little fertilizer at or near planting time can help jumpstart wheat toward a successful crop.
- ☆ Wheat is considered as highly responsive crop for certain fertilizers, particularly phosphorus and nitrogen.
- ☆ When applying a fertilizer for wheat, application methods and rates are much more flexible with phosphorus than nitrogen.
- ☆ Wheat plants normally show significant increase in falling of tillers and better root development with certain fertilizers like phosphorus and nitrogen.
- ☆ When applying fertilizer with seed, nitrogen and potassium rates should be limited to avoid potential toxicity to the seedling.
- ☆ When placing fertilizer in direct contact with wheat seed, wheat cultivators should apply following guidelines:
- ☆ In 15-inch spaced rows, use fertilizer not more than 14 kg of nitrogen witth potash for medium to fine textured soils or 9 kg for sandy or dry soils.
- ☆ In 10-inch rows, use fertilizer not more than 22 kg of nitrogen with potash for medium to fine soils and 12 kg for sandy or dry soils.
- ☆ For 6- to 8-inch rows, use fertilizer not more than 25 kg of nitrogen with potash for medium to fine soils and 18 kg for sandy or dry soils.

3.8 APPLY FERTILIZERS IN RIGHT DOSAGE, TIME AND INTERVALS OF CROP GROWTH

- ☆ Timing of fertilizer application has a significant effect on crop yields.
- ☆ Proper timing of the fertilizer application increases yields, reduces nutrient losses, increases nutrient use efficiency and prevents damage to the environment.
- ☆ Applying fertilizers at wrong time might result in nutrient losses, waste of fertilizer and even damage to the crop.
- ☆ Plants need different nutrient rates and ratios at different growth stages.
- ☆ To make the nutrients available when plant needs them, fertilizers should be applied at the right timing.

- ☆ The optimum timing for fertilizer application is determined by Nutrient Uptake Pattern of the crop.
- ☆ For similar crop, each nutrient has an individual uptake pattern.
- ☆ The maximum rate of fertilizer that can be applied at one application depends on the salinity threshold that the crop can tolerate.
- ☆ Split fertilizer applications helps in avoiding salt damages to the crop and improving germination rate.
- ☆ Soil type affects the timing and frequency of fertilizer application where Cation Exchange Capacity and Soil Texture properties determine the frequency and timing of application.
- ☆ If nitrogen is applied too early before need of plant, then a required portion may get lost before the crop takes it up, so time nitrogen is in soil before the plant takes up should be minimized.

Chapter 4

INTEGRATED WEED MANAGEMENT

4.1 WEEDS AND ITS TYPES

4.1.1 Introduction

- Weed is a plant considered undesirable in a particular situation.
- It includes plants unwanted in human-controlled settings like farm fields, gardens, lawns and parks.
- It has no botanical significance as a plant that is a weed in one context is not a weed when growing in a situation where it is in fact wanted, and where one species of plant is a valuable crop plant, another species in the same genus might be a serious weed, such as a wild bramble growing among cultivated loganberries.
- Volunteer crops are regarded as weeds in a subsequent crop.
- Many plants that people widely regard as weeds also are intentionally grown in gardens and other cultivated settings, in which case they are sometimes called beneficial weeds.
- Weed applies to any plant that grows or reproduces aggressively, or is invasive outside its native habitat.

4.1.2 Features

- A number of weeds like dandelion and lamb's quarter, are edible with leaves or roots used for food or herbal medicine.

- ☆ It prevent pest insects from finding a crop as their presence disrupts the incidence of positive cues which pests use to locate their food.
- ☆ Weeds act as a living mulch and provides ground cover to reduces moisture loss and in preventing erosion.
- ☆ It improves soil fertility by bringing up the nutrients like calcium and nitrogen from deep in soil with tap root and clover hosts nitrogen-fixing bacteria in its roots, fertilizing the soil directly.
- ☆ The dandelion species breaks up the hardpan in overly cultivated fields and help crops to grow deeper root systems.
- ☆ Many garden flowers originates as weeds in cultivated fields and have selectively bred for garden worthy flowers or foliage.

4.1.3 Types of Weeds

There are three types of common weed plants in regards to their growing characteristics which include:

Annual types

- ☆ It germinates and spread by seed having average lifespan of less than an year.
- ☆ It includes winter and summer types.
- ☆ It grows from seed, develop into mature plant, set flowers and seeds and die after seeds gets shed.
- ☆ They maintain their populations by production of seeds with each generation.
- ☆ Winter annuals like chickweed, germinate in late summer/early fall, go dormant in winter and actively grow during spring.
- ☆ Summer annuals like lambs quarters, germinate in spring, grow throughout summer and goes with the arrival of cold weather.

Biennial types

- ☆ It completes their life cycle in two years, germinates and forms rosettes in first year and produce flowers and seeds in second year.

- ☆ These weeds are best controlled in young growing stages of first-year plant.
- ☆ It includes bull thistle and garlic mustard.

Perennial types

- ☆ These plants are able to live for two years or more.
- ☆ It returns every year and produces long tap roots with seeds.
- ☆ It includes dandelions, plantain and purple loosestrife.
- ☆ It is most difficult to control.

4.1.4 Weeds in Wheat

- ☆ The main weeds in wheat crop are:

Name	Hindi Name	English Name	Type
Phalaris minor	Gahu Ka Mama	Little Seed Canary Grass	Grassy
Phalaris arundinacea	Gahunsa	Canary Grass	Grassy
Avana fatua	Janglee Jai	Wild Oat	Broadleaf
Convolvulus arvensis	Hiran khuri	Field Bind Weed	Broadleaf
Anagallis arvensis	Krishna Neel or Billi Booti	Scarlet pimpernal	Broadleaf
Fumaria parviflora	Ban Soya or Gajari	Fumitory	Grassy
Melilotus alba	Safed Senji or Khandi or Wild Sengi or Murela	White Sweet Clover	Broadleaf
Heliotropium ocalifolium	Sengi	Clover	Broadleaf
Launea mudicaulis	Jangli Gobhi	NA	Broadleaf
Euphorbia dracunculoides	Kaurgandal	Spurge	Broadleaf
Vicia hirsute	Mun Muna or Rari or Rewari	Vetch	Broadleaf

Name	Hindi Name	English Name	Type
Spergula arvensis	Jangali Bania or Sath Gathia	Corn Spurry	Broadleaf
Carthamus oxycantha	Pohli	Wild Safflower	Broadleaf
Argemone mexicana	Satya Nasi or Kateli	Mexican Poppy or Prickly Poppy	Broadleaf
Chenopodium album	Bathua	Lambsquarter or Dog's Tooth Grass	Broadleaf
Chenopodium murale	Bathua or Chenopodium	Nettle Leaf Goose Foot	Broadleaf
Euphorbia hirta	Badi Dudi	Pill Pod Spurge	Broadleaf
Asphodelus tenuifolius	Pyazi	Wild Onion	Broadleaf
Parthenium Spp.	Gajar Ghas	Wild Carrot Grass	Broadleaf
Achryanthus aspera	Chir Chita or Latzeera	Snake's Tail or Prickly Chaff Flower	Broadleaf
Cynodon dactylon	Dooba	Bermuda Grass	Grassy

4.1.5 Weeds causing damage to Wheat Crop

Weeds do enormous damage to wheat crop which includes:

- ☆ Depletion of soil nutrients and moisture and compete with crop plants for light and space thereby reducing crop yields.
- ☆ Damaging to crops harbouring pests and disease agents.
- ☆ Damaging of 10 to 40% wheat crop depending upon their intensity.

4.1.6 Stages of weed competition with field crop

Competition between weeds and crops is expressed by altered growth and development of both species. Interspecific competition occurs when two or more species coexist in time and space and simultaneously demand a limited resource. Intraspecific competition occurs when two or more plants of the same species coexist in time and space and simultaneously demand a limited resource.

- ☆ Period of weed competition is shortest time span during the crop growth when weeding results in highest economic returns.
- ☆ Crop yield level obtained by weeding during the period is similar to that obtained by the full seasons weed free conditions.

- ☆ Critical period of weed competition is period between early growth during which weeds can grow without affecting crop yield and point after which weed growth does not affect the yield.
- ☆ Critical period of weed competition is 1/3rd of period of the crop.

4.1.7 Factors affecting weed crop competition

Competition depends on certain interrelated factors such as:

Timing of weed emergence

- ☆ The first plant that effectively obtains water, nutrients and sunlight from a site and becomes established at that site has distinct competitive advantages over plants that develop later.
- ☆ The effect of a weed competition is greatest when the crop is young, since this is the stage which plant growth is inhibited most by inadequate light, water and nutrients.
- ☆ Crop yields are much more reduced by early season weed competition than by later season competition.

Growth form

- ☆ Growth form is manifested in two major parts:
- ☆ Growth habit: Extent of root development, height, leaf area, amount of branching
- ☆ Growth rate: Those which can develop canopy very rapidly over another, has definite advantage of shading over the second plant communities.

Weed Density

- ☆ The numerical superiority that weeds exhibit greatly reduces the availability of water, nutrients and light to crop plants and accounts for much of what we consider to be weed competition.
- ☆ Increase in crop population density distributes available resources among the crop community, but increase in weed population diverts available resources from the crop communities.
- ☆ Weed density is higher in distributed or agricultural soil than in undistributed soils.

Duration of weed growth

- ☆ The duration of weed growth is equally important with all other factors.
- ☆ If weeds are allowed to grow for an extended period crop yield may be drastically reduced.

- ☆ Weeds that are not controlled within 2-3 weeks of emergence usually affect crop yield.
- ☆ It is particularly important for upland rainfed crops like aus rice, jute etc.
- ☆ In many crops, weed infestation during the first 3-8 weeks is very critical, so crop fields must be kept weed free during this period.

Characteristics of Weed species

- ☆ Weeds differ in their ability to compete with crops at similar density levels due to differences in growth habits, allelopathic effect on germination, growth of neighbouring crop plants.
- ☆ Wild mustard reduced sugarbeet yield more than Setaria glauca (Yelow foxtail).
- ☆ In dry areas perennial weeds like Cirsium arvense (Canada thistle) and Convolvulus arvensis (Bindweed) are more competitive than annual weed species as of their deep roots and early, heavy shoots growth.

Characteristics of crop species

- ☆ Crops and their varieties differ in their competing ability with weeds.
- ☆ Among winter grasses, the decreasing order of weed competing ability includes barley, rye, wheat and oat.
- ☆ High tolerance of barley to competition from weeds is assigned to its ability to develop more extensive roots during its initial three weeks growth period than the other grains.

4.1.8 Positions along the field where the weeds are likely to grow

- ☆ In farming, farmers grew something due to soil property and pre available micro seeds shows growth of weeds which spoils the actual outcome of farming, since weeds affect the growth of planted plants.
- ☆ Weed detection is problem of accurately identifying the area of weeds so that specific areas can be targeted for spraying with minimum spraying on the other plants of interest.
- ☆ The weeds that are under target can easily, rapidly and accurately separated from the background.
- ☆ Once the green plant gets separated from soil, the next step is to extract inter-row weeds from plant pixels.

4.2 WEED CONTROL METHODS

Controlling weeds in crop production involves wide range of techniques which includes:

Preventative Weed Control

- ☆ It refers to control method that aims to prevent weeds from being established in a cultivated crop, a pasture, or a greenhouse.
- ☆ Examples include use of certified weed free seed, transporting weed free hay, making sure farm equipment is cleaned before moving from one location to another, and screening irrigation water to prevent weed seeds from traveling along irrigation ditches.

Cultural

- ☆ It refers to a technique that involves maintaining field conditions such that weeds are less likely to become established and/or increase in number.
- ☆ Examples of cultural weed control would be crop rotation, avoiding overgrazing of pastures or rangeland, using well-adapted competitive forage species, and maintaining good soil fertility.

Mechanical

- ☆ It refers to technique that involves the use of farm equipment to control weeds.
- ☆ The two mechanical control techniques most often used are tillage and mowing.

Biological

- ☆ It refers to technique that involves the use of natural enemies of weed plants to control the germination of weed seeds or the spread of established plants.
- ☆ It is rapidly expanding area of weed control with many examples which includes sheep to control tansy ragwort or leafy spurge, cinnabar moth and the tansy flea beetle to control tansy ragwort, the chrysolira beetle to control St. John's Wort, and the use of goats to control brush on rangeland.

Chemical

- ☆ It refers to technique that involves the application of a chemical (herbicide) to weeds or soil to control the germination or growth of the weed species.
- ☆ It includes chemicals to control weeds which includes 2,4-DB; EPTC; bromoxynil; and paraquat.

4.3 KNOWLEDGE ABOUT THE CHEMICALS/HERBICIDES

- ☆ Herbicides play a vital role in integrated weed management programs.
- ☆ Knowledge of the mechanisms and activity of herbicides will improve the impact and sustainability of herbicides as a weed management tactic.

4.3.1 Types of herbicides

Translocated herbicides:

- ☆ They move to the site of action by transport mechanisms in plant; xylem and phloem.
- ☆ The xylem transports water and nutrients from the soil to growth sites and phloem transports products of photosynthesis (for instance, sugars) to growth and storage sites.
- ☆ It may take up to two weeks for symptoms to develop on the target weeds depending on herbicide rate, conditions and species.

Contact herbicides

- ☆ They have limited movement within the plant, so complete coverage of target is critical.
- ☆ Compared to translocated herbicides (glyphosate), contact herbicides (paraquat, oxyfluorfen, diquat and bromoxynil) tend to show symptoms rapidly, usually within 24 hours.

Selective herbicides

- ☆ It kill target weeds and not desired plants (crop or pasture) when applied at specified application rate.
- ☆ **Non-selective herbicides**
- ☆ It is also called as knockdown herbicides like glyphosate or paraquat that damage most plants.

Residual herbicides

- ☆ It remains active in the soil for an extended period of time and acts on successive weed germinations.

Non-residual herbicides

- ☆ It includes non-selective paraquat and glyphosate.
- ☆ It has little or no soil activity and are quickly deactivated in the soil.

- ☆ They are broken down or bound to soil particles and is less available to growing plants.
- ☆ They have little or no ability to get absorbed by roots.

Post-emergent and pre-emergent

- ☆ These are target and timing of herbicide application.
- ☆ Post-emergent is foliar application of the herbicide after the target weeds have emerged from the soil.
- ☆ Pre-emergent is the application of herbicide to the soil before weeds have emerged.
- ☆ Herbicide mixtures and sequential applications
- ☆ It involves application of more than one herbicide to increase spectrum of weed species controlled for resistance management.
- ☆ A mixture involves the application of multiple products in a single application.
- ☆ Where herbicides are antagonistic and cannot be mixed together in a single tank, they are applied sequentially.

4.3.2 Classification of herbicides based on application methods

Soil application

Soil surface application

- ☆ Herbicides are usually applied to soil surface to form a uniform herbicide layer.
- ☆ Applied herbicides, due to their low solubility may penetrate only few centimetres into the soil.
- ☆ Weeds germinating in the top layers are killed due to incidental absorption of herbicides.eg. triazines, ureas and anilide

Soil incorporation

- ☆ Many herbicides are applied to soil surface and incorporated into the soil either by tillage or irrigation for their effectiveness.eg. volatile herbicides viz., aniline and carbamate

Sub surface application

- ☆ Perennial weeds Cyperus rotundus and Cynodon dactylon are controlled by injecting herbicides to the lower layers of the soil at several points.

Band application

- Herbicides are applied as narrow bands over or along the crop row.
- Weeds in between crop rows can be controlled by inter-cultivation or band application of herbicide.
- It is useful method where labour is expensive and inter-cultivation is possible. eg. weeds in maize can be controlled effectively by spraying atrazine on seed row at the time of sowing.

Foliar application

Blanket application

- It is applied over the entire leaf area.
- Selective herbicides are applied by this method.

Directed application

- Herbicide is applied directly to weeds between crop rows, avoiding the crop foliage.
- Care is taken to avoid spray fluid falling on the crop. eg. late weeds in cotton can be controlled by spraying non selective herbicide by directed spray.

Spot application

- Herbicides are applied or poured on small patches of weeds, leaving the relatively weed free patches untreated.
- It minimizes the herbicide usage per unit area.

4.4 INFORMATION ABOUT THE CHEMICALS/HERBICIDES EFFECTIVE IN CONTROLLING THE WEEDS

- Chemicals that are used to kill plants or weeds are called herbicides.
- Herbicide selectivity and its dose, time and method of application are of paramount importance/consideration before applying to a crop.
- In chemical weed control, chemicals called herbicides are used to kill certain plants or inhibit their growth.
- Chemical weed control is an option in integrated weed management that refers to the integrated use of cultural, manual, mechanical and/or chemical control methods.

4.4.1 Control of Weeds

- ☆ Prevent yield loss due to weed competition
- ☆ Maintain purity and/or quality and market price of harvested grain
- ☆ Prevent build-up of weed seeds in soil
- ☆ Prevent weeds that may attract insects or rodents (rats) or act as a host for diseases
- ☆ Prevent clogging of field irrigation channels and facilitate water flow
- ☆ Reduce time and cost of land preparation and weeding operations

4.4.2 Use of Herbicides to Control Weeds

Herbicides are used to control the weeds due to:

- ☆ Less labor (0.5 person-day per ha per application) and less drudgery
- ☆ Cost-effective, if practiced properly
- ☆ No need to wait for weeds to grow bigger for hand weeding
- ☆ Herbicides can differentiate between rice and weeds, even at seedling stage where it is very difficult for people to see the difference

4.5 PRE EMERGENCE WEED CONTROL

4.5.1 Introduction

- ☆ In India, main weeds in winter grain crops are Board leaved weeds such as Lamba quarters (Chenopodium sp.) and Wild safflower (Carthomus Osycantha)
- ☆ In some parts of the country, perennial dicot weeds such as canda thistle (Cirsium avenses), field bind weed (Convalvelus arvensis) are found.
- ☆ Annual Grasses like Wild Oat (Avenu fatua), Chanry grass (Philaris minor).
- ☆ The weed seeds get admixture in the crop seed and reduce grain quality.

4.5.2 Pre-emergence weed killers

- ☆ These are used on established fields as part of annual effort to combat plant pests.
- ☆ These chemical compositions are used before weeds take hold to kill off infant root systems and keep them from growing.
- ☆ It is used before seeing weeds to save them from showing up in fields which has no interference of chemicals with germination but can save formation of new root cells in baby weed plants.
- ☆ Timing, weather and the type of weeds that are problematic in fields will dictate the exact formula and application for using pre-emergence procedures.
- ☆ Chemicals used in pre-emergence are not effective on vegetative buds that sprout from existing roots or rhizomes.
- ☆ The chemicals cannot be used on prepared grass seed bed as their root stunting action in young plants also affects sprouting grass.
- ☆ Grown crop has no fear, as their root system is already developed and the crop is hearty and healthy.
- ☆ Pre-emergent procedure shows that it is the sensitive root tissue of newly germinated seedlings which is killed off, that result in complete plant death.
- ☆ Perennial weeds develop thick persistent adult roots that re-sprout in spring, which makes them difficult to control with pre-emergent formula.
- ☆ Timing of pre-emergence weed killer should match with germination period for the variety of weed.
- ☆ Biennial weeds are not controlled by pre-emergence procedure as they produce seed that germinates nearly year around.
- ☆ These weed killers need water to activate themselves and carry the chemical down to the root systems of newly sprouted weeds.

4.5.3 Pre-emergence herbicide use on wheat

- ☆ Herbicides are usually applied immediately after planting.
- ☆ In pre-emergence, the herbicide is applied after the crop seeds are planted but before the crop and weeds have emerged from the soil.
- ☆ In this, the group of herbicides usually requires a rain within a few days after application to move the herbicide from the soil surface down into the top layer of the soil where most weed seeds are located.

- ☆ In pre-emergence, the herbicide includes Lasso, Dual, Lorox, A Atrex and B ladex.
- ☆ Pre-emergence treatments in wheat can be inconsistent in effectiveness as they require rainfall to be activated.
- ☆ If weeds and cheat grasses emerge before the herbicide is activated, control may be poor, especially the grasses.
- ☆ If there is enough rain to activate the herbicide before weeds and cheat grasses emerge, control or suppression will result good.
- ☆ The herbicides labeled for pre-emergence applications will be most consistent when applied as fall post emergence treatments.
- ☆ Labels of these herbicides differ in what is allowed with a pre-emergence application.
- ☆ Finesse allows for high use rate when applied as pre-emergence than as post emergence treatment for good season as long control of susceptible broadleaf weeds, till they are resistant.

4.5.4 Spray techniques: Procedure and precautions

- ☆ In annual crop application after the sowing of the crop and before the emergence of weeds shows pre-emergence to the crop.
- ☆ In soil application by spraying of atrazine on 3rd DAT can be termed as pre-emergence to crop while soil application by spraying the immediately after a rain to control a new flush of weeds in inter-cultivated orchard can be specified as pre-emergence to weed.
- ☆ Eg. Atrazine, Pendimethalin, Butachlor, Thiobencarb, Pretilachlor

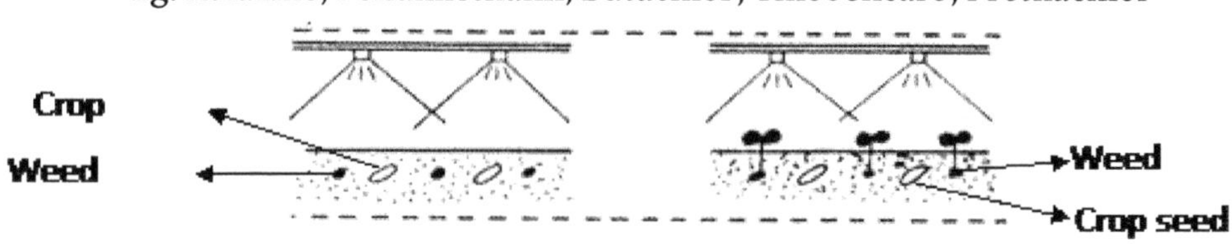

- ☆ There are three spraying process;

Target:

- ☆ It is critical to ensure that the product selected is the most appropriate for weeds present and the situation.
- ☆ It is important to consider factors like previous weed burden, potential resistance, products residual activity, solubility and soil moisture requirements, ability to tank mix, and target area including stubble load and ground cover.

- ☆ Knowing about product labels and refer manufacturer technical data for taking any informed decision.

Timing:

- ☆ It is important to consider the appropriate conditions for spraying pre-emergent and know the timing of incorporation to be done.
- ☆ Suitable conditions for spraying pre-emergent products are the same as for any spraying operation.
- ☆ Extremes of temperature, relative humidity and wind speed all effect the amount of product available to deposit onto the target area which should interact with droplet size, speed and boom height.
- ☆ Refer to product label for required spraying conditions and should follow guidelines:
    - Measure and record conditions at the site of application, before, during and at end of days spraying operations.
    - Never spray when the wind speed is less than 3 km/h and avoid spray when wind speeds exceed an average of 15 km/h.
    - Do not spray when the wind is blowing towards a sensitive area (including fallows to be planted to different crops) unless you have a sufficient buffer in place (typically 300m or more when using a coarse spray quality).
    - Never spray during highly stable conditions like surface temperature inversion conditions
    - Carefully monitor the Delta T value (temperature and relative humidity).
    - When Delta T exceeds 10 it is probably time to stop spraying. The only situation where it may be reasonable to continue would be when you are using a spray quality that is larger than coarse (very coarse or extremely coarse), and you have adequate soil moisture and ground cover, such that the delta T value at the target (soil level) is below 8.
    - Continue to monitor and record factors such as rainfall after the application.

Technique:

- ☆ Many small adjustments to the sprayer setup can improve the deposit onto the soil and minimise losses, such as tie up on stubble and loss to atmosphere.
- ☆ The heavier the stubble load, more will coarse spray quality or larger droplets.
- ☆ Larger droplets usually require greater application volumes (more than 80 L/ha) to ensure uniformity of the deposit for pre-emergent herbicides with low solubility in the soil.
- ☆ When using solid droplets in light stubbles a medium to coarse spray quality may be appropriate, provided the conditions are suitable.
- ☆ If using medium spray quality expect that a high proportion of droplets will be retained on stubble, and risk of spray drift and loss to atmosphere will increase. For other pre-emergent herbicides that bind strongly with organic matter, retention on the stubble means a tie up of the product, unless the stubble was completely buried during incorporation.
- ☆ In heavy stubbles, penetration is critical as with non air induction nozzles, use coarse to very coarse spray qualities. Larger solid droplets tend to bounce off the standing stubble onto the soil surface.
- ☆ If travelling at lower speeds (around 16 km/h or less) a twin fan nozzle, such as the TTJ may be a good choice to improve penetration and increase uniformity.
- ☆ At speeds above 16-20km/h a twin fan becomes less suitable, particularly for medium to coarse spray qualities, where nozzles producing a standard fan pattern would be more appropriate.
- ☆ At higher speeds (above 16km/h) it is good to use nozzles which are slightly offset (usually angled backwards).
- ☆ If the nozzle spacing is different to row width, ideal height above the target is at a point where double overlap is achieved at top of stubble, so narrow nozzle spacing allows reduced boom heights above the target.

4.5.5 Precautions and procedure to control pre emergence weeds with appropriate herbicides

Preemergence Herbicides

- ☆ Atrazine (Purge II – 2 lbs/gallon, Atrazine Plus – 1.2 lbs/gallon, Aatrex 4L)
- ☆ Benefin (Balan 2.5G)

- ☆ Benefin + oryzalin (XL 2G)
- ☆ DCPA (Dacthal 75W) and (Weed Killer Granules 5G)
- ☆ Dithiopyr (Dimension 1EC, 2EW, 40WP, others)
- ☆ Indaziflam (Specticle 20WSP, Specticle Flo)
- ☆ Oryzalin (Surflan 4AS, others)
- ☆ Pendimethalin (Halts 1.71G, Pendulum 2G, Pendulum Aqua Cap 3.8CS, Pendulum 3.3EC)
- ☆ Prodiamine (Barricade 4FL, Barricade 65WG, Cavalcade 65WDG, Stonewall 65WDG, Regalkade 0.5G)
- ☆ Simazine (Princep 4L, WynStar, others)

Weed Control Safety/Precautions

Herbicides can be effective in controlling weeds, but be careful to handle these powerful chemicals properly and safely.

1. If Leader/SF-10 herbicide is used, do not sow sorghum (jowar) and maize crops in the following season.
2. Spray to be done with patience and is uniform as required.
3. For spray of Leader/SF-1 0/ Topik / Point / Puma PDWer, 100 litres of water per acre is sufficient.
4. PBW 343 can tolerate 2, 4-D very well provided it is sprayed at maximum tillering stage.
5. Do not use Leader/SF-10 where any cruciferous crop has been sown as these crops are highly sensitive to herbicide. In such case, Isoproturon group of herbicides or Puma PDWer or Topik / Point needs to be used where raya and sarson shows enough tolerance.
6. Spray on clear days where after spray, first and second irrigation should be light. Flooding or heavy irrigation reduces the effect of herbicides.
7. In broad-leaf weeds, problem with Phalaris, mix is recommended for dose of Isoproturon with 2, 4-D sodium salt at 250 g/acre or 2, 4-D ethyl ester (36%) at 250 ml / acre in 200 litres of water and use knapsack sprayer when crop is at maximum tillering stage (35-45 days after sowing).
8. Dissolve each herbicide in some water separately and mix both the solutions in desired quantity of water.
9. Prefer use of 2, 4-D sodium salt as mixture with 2,4-D should not be applied when wheat is intercropped with rayal sarson gobhi sarson or gram.
10. Joint infestation of Gulli danda / wild oats and broad leaf weeds in wheat can be controlled using tank mix application of Topik / Point / Moolah /

Puma PDWer with 2, 4-D / Algrip. All these herbicides are applied at their full-recommended rates, 30-35 days after sowing of crop.

11. Spray required dose of Stomp or Treflan as pre- emergence in 200-300 litres of water per acre on well prepared seedbed, free of clods and plant debris.
12. Good soil moisture and high volume spray are key to success for both herbicides.

4.5.6 Advantages of Pre-emergence Herbicides

Some of the advantages of using pre-emergence herbicides on wheat crop will result as:

- ☆ Pre-emergence herbicides are applied prior to weed seed germination and emergence.
- ☆ Many pre-emergence herbicides are available in granular formulations where granules are easier to apply than spray formulations. Also, granules are not susceptible to spray or vapor drift that can occur with spray formulations.
- ☆ With exception of atrazine (Aatrex, Purge II, Atrazine Plus), many ornamental shrubs, trees and flowers are tolerant of pre-emergence herbicides. The, DCPA (Dacthal), oryzalin (Surflan), benefin (Balan) and benefin + oryzalin (XL) are labeled for use on selected ornamentals.
- ☆ Pre-emergence herbicides do not cause injury to established turfgrasses.

4.6 POST EMERGENCE WEED CONTROL

4.6.1 Introduction

- ☆ Post-emergent's attack the weeds once they grow up.
- ☆ Post-emergence herbicide controls the existing weeds and helps in preventing future ones.
- ☆ There are different types of post-emergent weed killers, what cultivator should keep information related to variety of weeds and need to control them.
- ☆ Post-emergent weed killers attack the foliage or flow systemically to the roots of the weed.
- ☆ Herbicide application after the emergence of crop or weed is referred as post-emergence application.
- ☆ When the weeds grow before the crop plants have emerged through the soil and are killed with a herbicide then it is called as early post-emergence.

- ☆ Spraying 2,4-D Na salt to control parasitic weed striga is called as post-emergence while spraying of paraquat to control emerged weeds after 10-15 days after plantation can be called as early post-emergence.
- ☆ Eg. Glyphosate, Paraquat, 2,4-D Na Salt.

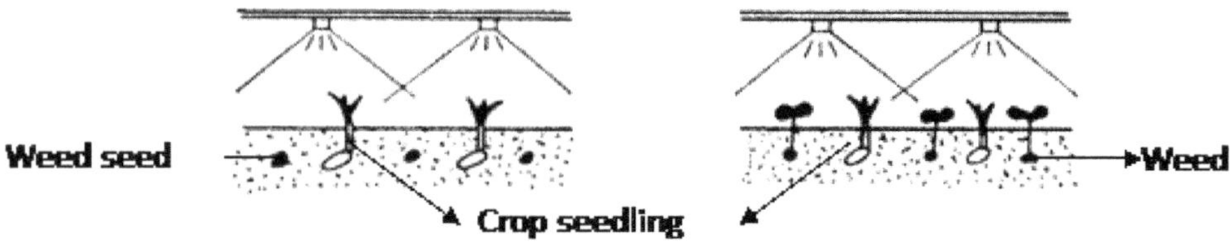

4.6.2 Types of Post-Emergent Weed Killers

- ☆ The post-emergent can be either systemic or contact applications.
- ☆ Systemics are useful on perennial weeds as they are absorbed directly in plant and move throughout it for maximum killing action.
- ☆ Contact herbicides kill the exposed part of plant and are used on annuals and smaller weeds which kill the entire plant.

4.6.3 Post emergence herbicides

- ☆ These herbicides are applied directly to emerge weeds.
- ☆ In contrast to pre-emergence herbicides, post emergency herbicides controls only weeds that are emerged and actively growing at the time of treatment.
- ☆ They are used for farm weed control which is systematic or translocated in the plant system that transports food and water.
- ☆ It is applied after the weeds and crop have emerged from the soil.
- ☆ The treatments can be applied in either a broadcast or directed fashion.
- ☆ When applying post emergence herbicides, it is necessary to have maximum coverage of the weed with the spray solution.
- ☆ Surfactants are used with post emergence herbicides to enhance control.
- ☆ These herbicides are classified as selective and non-selective.
- ☆ Selective herbicides are used to target certain weeds, and in areas such as turf where contact with the grass is unavoidable.
- ☆ Non-selective herbicides are used for broad weed control and have a purpose in open, unmanaged fields, for instance.
- ☆ They control weeds that have emerged since crop or pasture establishment and can be applied with little damage to the crop or pasture plants.

Benefits

- ☆ It gives high level of target weed control with extra benefit of improved crop or pasture yield.
- ☆ It allows fine-tuning of herbicide selection to match weeds present in the paddock.
- ☆ In this, timing of application is flexible to suit weed size, crop growth stage and environmental conditions.
- ☆ It has pre-emergent activity on subsequent weed germinations.

Issues

- ☆ It requires careful consideration when selecting the best post-emergent herbicide in any one situation.
- ☆ Use of post-emergent herbicides to stressed crops and weeds results in reduced levels of weed control and increased crop damage.
- ☆ In this, crop competition is important for effective weed control by selective post-emergent herbicides.
- ☆ It requires suitable technique for the situation to optimise control.
- ☆ It requires correct adjuvant to ensure effective weed control.
- ☆ It needs suitable formulation of herbicide for every situation.
- ☆ The effectiveness of post-emergent herbicides is influenced by range of plant and environmental factors.

4.6.4 Mechanical control: Application of mechanical weeding equipments such as hoes etc.

- ☆ Mechanical weed control allow wheat cultivator to reduce or even eliminate herbicide use, and contribute to better environment.
- ☆ Mechanical weeding is opposite as it aerates the soil, therefore stimulating crop growth.
- ☆ In mechanical weeding, weeds are destroyed when buried by moving soil.
- ☆ Tools such as rotary hoe or finger weeder, used at pre-emergence and/or post-emergence of the crop are good.
- ☆ Mechanical weeding is used for spring and winter cereals in any type of soil except organic soil.
- ☆ Mechanical weeding is effective against many annual weeds like wild mustard, lambs quarter, annual grasses, etc.

☆ It is effective against young biennial seedlings like dandelion.

Mechanical weeding equipments

1. Rotary Hoe

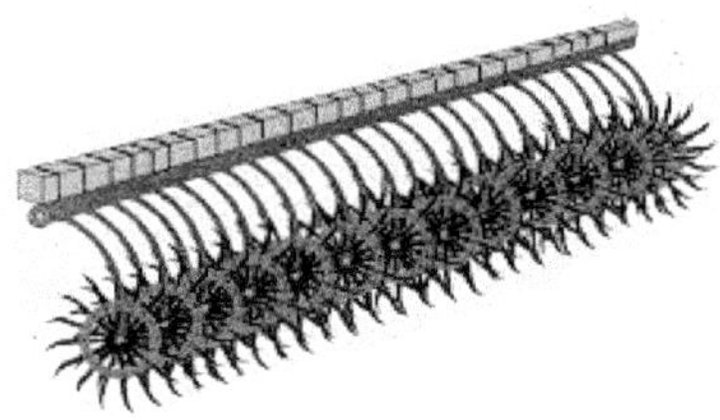

☆ It has dozens of closely adjacent wheels which works by throwing the soil as it rolls shallowly over the field.

☆ In this, many models are for high residue conditions like those found under minimum tillage systems.

☆ It has mild impact on the soil surface.

☆ It is important for use when the weeds are just coming through the soil or starting to emerge.

☆ In this, last weed stage lasts 4 to 8 days as per the climate.

☆ Here, stage of the crop is not as critical as for the finger weeder as the rotary hoe has more gentle action.

☆ It gives best results in slightly dry and crusted soils and also in sandy soil if it is little moist.

Weeding strategy

Crop Stage	Weeding
Pre-emergence	Almost obligatory, 3 to 7 days after sowing at a speed of 12-20 km/h.
1 leaf	6-8 km/h and only for high weed populations
2 to 5 leaves	12-30 days after sowing at a speed of 10-14 km/h. Crop will look terrible but yields will not be reduced and is done, even if, weed pressure is not high, to stimulate crop growth and possibly increase yield.

Adjusting rotary hoe

☆ Use the rotary hoe for about hundred meters.

☆ Check the crop making as less than 10% of plants are affected.

- ☆ Partly buried plants are not considered damaged.
- ☆ Verify 80-90% of the weeds at cotyledon stage which are buried and that whole surface is disturbed by comparing it with adjacent non-weeded surface.

2. Finger Weeder

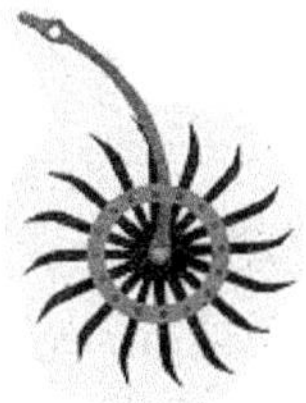

- ☆ It has long tines that vibrate in the soil.
- ☆ It is made of several sections that allow implement to follow the surface of the field.
- ☆ It has a lever that allows adjustment of all tines in a section at once.
- ☆ Depth wheels allow better precision when adjusting.
- ☆ In this, crop stage is critical due to its stronger action.
- ☆ The finger weeder is not good if lot of residue is present on the soil surface.
- ☆ It will drag the residue and damage the crop.
- ☆ Its effectiveness is reduced after first leaf-stage of weeds.

Crop Stage	Weeding
Pre-emergence	10-15 km/h. 3 to 7 days after sowing when white threads are present.
2 to 3 leaves	5-6 km/h, if weed pressure is high and plants have two leaves. Set finger pressure at minimum.
3 to 5 leaves	6-9 km/h. 15-30 days after sowing. Crop will look terrible but yields will not be reduced. Do not go over twice. May be done, even if weed pressure is not high, to stimulate crop growth and possibly increase yield.

Adjustment

- ☆ If there is multi-section finger weeder, adjust the tines of each section at different pressure.
- ☆ Adjust the depth wheels so that all tines are slightly in the soil with supporting chains well extended.

- ☆ Adjust the three point hitch so that the sections are horizontal.
- ☆ Drive for a small distance by noting its speed.
- ☆ Stop and check that less than 10% of the crop seedlings are damaged.
- ☆ Check that 80-90% of the weeds are buried and that entire surface is disturbed.

3. Tine weeders

- ☆ It has multiple rows of flexible metal tines that cover the entire soil surface.
- ☆ It wiggle slightly when pulled along, uprooting or dislodging small weeds.
- ☆ It works good at relatively high speeds.
- ☆ Its wide span makes the cultivation quick.
- ☆ It is rear-mounted and is available in many widths.
- ☆ In this, the tines tension can be adjusted on certain units or by pressure on 3-point hitch on other units.
- ☆ The gauge wheels are used to maintain uniform depth.
- ☆ It works best in friable soils which are free of rocks and also when weather allows weeds to dry out on soil surface after uprooting.

4. Basket weeders

- ☆ It is of relatively high-speed, between-row cultivators.
- ☆ It is good for control of small weeds in narrow rows when crop is small.
- ☆ The wire baskets work the soil surface and do not move soil into the row.
- ☆ It will not work well in crusted or rocky soil.
- ☆ The front set of baskets are ground driven where rear set of baskets are turned faster by chain and gearing, causing scuffing of the soil.
- ☆ It is available in 2 to 6 row units which are rear or belly-mounted.

5. Row crop cultivator

- ☆ It has gangs of S-shanks with sweeps that break up the soil by destroying small to moderate-sized weeds.
- ☆ The gangs work behind the tractor between the rows at uniform intervals.
- ☆ In this, shields are added to protect the crop from moving soil.
- ☆ In this, fertilizer hopper is used for side dressing during cultivation.

6. Rolling cultivator

- ☆ It has gangs of soil driven spider wheels which is kept independently on toolbar.

- ☆ It has an angle which works the soil with aggressiveness that can be adjusted.
- ☆ In this, numbers of gangs are grouped together which shows cultivator widths that are rear-mounted.
- ☆ In this, pairs of gangs are belly mounted to work a row or two.
- ☆ It throws the soil in row to bury small weeds or to form hills as per the angle of gangs.
- ☆ It is a relative heavy and aggressive tool.

7. Wiggle hoe

- ☆ It is in-row cultivator which uses hand steered side knives that is controlled by operator who moves it, in and out of the row.
- ☆ In this, the knives reach close to the row and remove the weeds, but when operator is careless, the crop can get damaged.
- ☆ It is good for irregular plant spacing in the row or between the rows.

8. Reigi weeder

- ☆ It is a new tool that needs a rear operator who steers a pair of rotating horizontal wheels in and out of crop row.
- ☆ In this, wheels are turned by PTO-driven belt having stiff tines that root out the weeds.
- ☆ It has wheels which come in many sizes for different row spacing's while the units come in 1 or 2-row models.
- ☆ It is good for killing weeds in and next to the row in widely spaced crops.

4.6.5 Chemical control: Type of weeds (Broad/Narrow)

- ☆ Chemical weed control is an effective way to control weeds if applied at proper time. Weeds can be classified according to the type of their leaves:
- ☆ Broadleaf weeds
- ☆ Narrow-leaf weeds

Chemical control of broadleaf weeds

- ☆ Post emergence broadleaf herbicides are most effective when weeds are actively growing and when air temperatures is more than 70° F.
- ☆ Broadleaf weeds are controlled with selective post emergence herbicides such as 2,4-D, 2,4-DP (dichlorprop), MCPP (mecoprop), dicamba, clopyralid, and triclopyr.

- The most effective control of broadleaf weeds is obtained when post emergence herbicides are applied as sprays to foliage.
- Using amine salt in water, it can be sprayed as amine formulations are nonvolatile, as it can react with calcium and magnesium ions to form insoluble salts.
- Granular formulations of the products are used to control broadleaf weeds, so granular should be applied to moist foliage for optimum control.
- Broadleaf weeds can be controlled by adding the fast-acting herbicide, carfentrazone, to formulations having slow acting, systemic herbicides.
- Carfentrazone on broadleaf weeds will act by disrupting chlorophyll synthesis that results in rapid yellowing and desiccation.
- Clopyralid is broad-spectrum broadleaf herbicide in many herbicide formulations which is used widely in cool season.

4.6.6 Post emergence Herbicides

- 2,4-D + MCPP (Weed-B-Gon Ready Spray; Weed-B-Gon RTU)
- 2,4-D + MCPP + Dicamba (33 Plus, Ace Lawn Weed Killer)
- Atrazine (Purge II – 2 lbs/gallon; Atrazine Plus – 1.2 lbs/gallon)
- Carfentrazone (Quicksilver – 1.9 lbs/gallon)
- Fenoxaprop
- Foramsulfuron
- Halosulfuron
- Imazaquin (Image – 1.5 lbs/gallon)
- MCPP + 2,4-D + dicamba
- sethoxydim (Segment, Vantage, others)
- simazine (Princep 4L, WynStar 4L, others)
- sulfentrazone
- sulfosulfuron
- thiencarbazone + iodosulfuron + dicamba

Chapter 5

IRRIGATION MANAGEMENT

5.1 REQUIREMENT AND AVAILABILITY OF WATER

5.1.1 Need of Water for Crop

- ☆ Water is one of the most important inputs essential for production of crops.
- ☆ Plants need water continuously during their life in huge quantities.
- ☆ It influences photosynthesis, respiration, absorption, translocation and utilization of mineral nutrients, and cell division apart from other processes.
- ☆ Water shortage and excess will affect the growth and development of a plant directly and, consequently, its yield and quality.
- ☆ Water affects the performance of crops not only directly but also indirectly by influencing the availability of other nutrients, the timing of cultural operations, etc.
- ☆ Water and other production inputs interact with one another.
- ☆ In proper combinations, crop yields can be boosted manifold under irrigated agriculture.
- ☆ It is a costly input when canals supply it.
- ☆ The constructing of dams and reservoirs, the conveying of water from storage points to the fields, the operating and the maintaining of canal systems involve huge expense.

- ☆ The misuse of water leads to problems of water logging, salt imbalance, etc. which render agricultural lands unproductive.
- ☆ Proper appreciation of relationship and economic utilization of water resources is required for maximum crop production.

5.1.2 Water requirement for crops

- ☆ It is essential to know the water requirement of a crop which is the total quantity of water required from its sowing time up to harvest.
- ☆ Naturally different crops may have different water requirements at different places of the same country, depending upon the climate, type of soil, method of cultivation, effective rain etc.

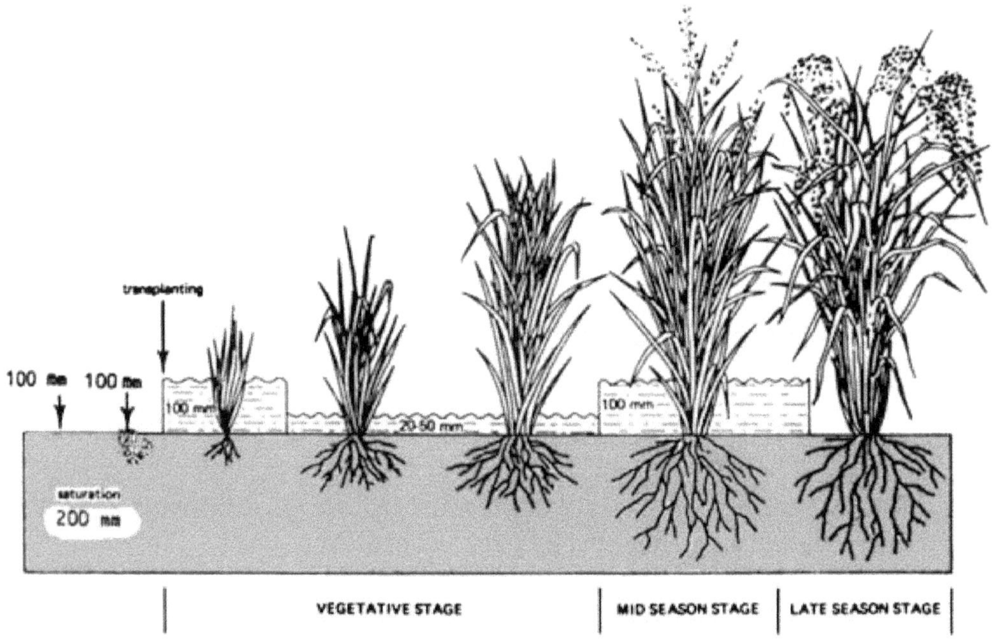

- ☆ The total water required for crop growth is not uniformly distributed over its entire life span known as crop period.
- ☆ The watering stops same time before harvesting and time duration from first irrigation at the time of sowing to last before harvest is known as base period.
- ☆ As water is essential for crop growth, it's requirement for each crop is definite which differs as per the season (Kharif, Rabi or Summer) or as per soil type (light, medium and heavy) but net requirement results same.

5.1.3 Total water requirement in growing crop

The water that is required to irrigate a field or plot of land growing the particular crop not only has to satisfy the evapo-transpiration needs for growing the crop, but also include:

- ☆ Losses in form of deep percolation while conveying water from inlet of the field till its last or tail end as water gets distributed within the field
- ☆ Water requirement for special operations like land preparation, transplanting, leaching of salts, etc.
- ☆ Evapo-transpiration requirement of crops will not include water required by crops for building up plant tissues, as compared to evaporation needs. ETC is often related as the consumptive irrigation requirement.

The net irrigation requirement is the amount of irrigation water required to be delivered in the field to meet the consumptive requirement of crop and other needs like:

- ☆ leaching,
- ☆ pre-sowing
- ☆ nursery water requirement

Thus, NIR (net irrigation requirement) = CIR + LR + PSR + NWR

Where

- ☆ LR = Leaching requirement
- ☆ PSR = Pre-sowing requirement
- ☆ NWR = Nursery water requirement

The total depth of water required to raise a crop over a unit area of land is known as delta which has some typical values for common crops such as:

Rice

- ☆ 1000mm to 1500mm for heavy soils or high water table
- ☆ 1500mm to 2000mm for medium soils
- ☆ 2000 to 2500 for light soils or deep water table
- ☆ 1600mm for upland conditions

Wheat

- ☆ 250mm to 400mm in northern India
- ☆ 500mm to 600mm in Central India

- ☆ Barley: 450mm

Maize

- ☆ 100mm during rainy season
- ☆ 500mm during winter season
- ☆ 900mm during summer season
- ☆ Cotton: 400 – 500mm

Sugarcane

- ☆ 1400mm to 1500mm in Bihar
- ☆ 1600mm to 1700mm in Andhra Pradesh
- ☆ 1700mm to 1800mm in Punjab
- ☆ 2200mm to 2400mm in Madhya Pradesh
- ☆ 2800mm to 3000mm in Maharashtra

5.1.4 Irrigation requirement of crops

Crop	Growing Period (days)	Total Water Requirement		Daily Water Requirement	
		in cm	in inches	in cm	in inches
Jawar	114	64.25	25.70	0.575	0.23
Maizc	100	44.50	17.80	0.450	0.18
Rice	93	104.50	41.80	1.075	0.43
Wheat	88	37.00	14.80	0.425	0.17
Groundnut	124	65.25	26.10	0.525	0.21
Linseed	88	31.71	12.68	0.350	0.14
Cotton	202	105.50	42.20	0.525	0.21
Sugarcane	365	237.50	95.00	0.650	0.26
Tobacco	132	98.00	39.20	0.750	0.30
Onion	120	75.00	30.00	0.625	0.25
Potato	88	30.00	12.00	0.750	0.30
Pea	88	30.00	12.00	0.350	0.14
Mustard	88	25.20	10.08	0.300	0.12
Barley	88	25.20	10.08	0.400	0.16
Oat	88	36.00	14.40	0.400	0.16
Ragi	127	74.50	29.80	0.575	0.23

5.1.5 Availability of water

- ☆ Water is one of the most important inputs essential crops.
- ☆ The water resource potential of the country has been assessed from time to time by different agencies.
- ☆ While the total water resource availability in the country remains constant, the per capita availability of water has been steadily declining since 1951 due to population growth.
- ☆ The twin indicators of water scarcity are per capita availability and storage.
- ☆ A per capita availability of less than 1700 cubic metres (m^3) is water-stressed condition while if per capita availability falls below 1000 m^3, it is termed as a water scarcity condition.
- ☆ On an average, water-stressed condition on individual river basin-wise situation shows that, 9 out of 20 river basins with 200 million populations are facing a water-scarcity condition.

5.1.6 Sources of water

There are two sources of water

- ☆ Canal or surface water
- ☆ Ground water-dug wells and bore wells

Canal water:

- ☆ The canal water is available from rivers after construction of dams and storing in dams.
- ☆ Rains in seasons causes availability of water and when there are no rains, the problem of water shortage or non-availability of water arises for irrigation as well as for drinking.
- ☆ The artificial storages are created by constructing dams across rivers.
- ☆ The rivers are small to big, so dams of various sizes and capacity are built which are classified in three categories
    - Minor-command less than 5000 ha
    - Medium-Command 5000 to 10,000ha
    - Major-Command more than 10,000 ha
- ☆ The command of minor dams is small, their benefits are accrued in short period.
- ☆ The designing, planning and construction of large dams take long time and there benefit is available after long period.

- ☆ Water from the dams is made available for drinking and other human purposes, for industries and for irrigation.
- ☆ Huge capital operating and administrative costs are involved in making dam water available for above-mentioned purposes.

Well Water:

- ☆ Well is traditional source of water.
- ☆ They are constructed by individual farmers from their own funds or bank loans as it does not require much technical know-how.
- ☆ Wells can be completed in a short period and their benefits are available immediately.
- ☆ The farmers have full control on the use of waste for crops which helps them in deciding what crops to grow, when to irrigate and how much water to apply.
- ☆ There is generally no misuse or excess use of well water and hence no spoilage of lands.
- ☆ The command of each well is very small (5 to 10 ha) but the number of wells is very large, so total area irrigated by wells is substantial.
- ☆ As farmers have control on use of water, delicate crops or crops requiring frequent watering but less water at a timc, are grown on well water.
- ☆ Fruit, vegetables and differences in cropping patterns under canal and flower crops are commonly grown on well water.
- ☆ There are differences in cropping patterns under canal and well irrigations.

Irrigation Layouts:

- ☆ The water requirements of different crops are different with their root systems and feeding zones different along with different soil types.
- ☆ The following types of layouts are used for irrigating crops such as:
  - Ridges & furrows - e.g. Sugarcane and different types of vegetables, chillies.
  - Long beds (Saras) - e.g. Wheat, Jowar, Bajra, Groundnut, Maize, etc.
  - Small beds - e.g. Onion, Potato.
  - Rings - e.g. Gourds (Vine crops)
  - Raised beds - e.g. Raising seedlings of various crops.
  - Trenches - e.g. grapes, banana

5.2 CRITICAL STAGES OF CROP GROWTH FOR IRRIGATIONS

- ☆ Most favorable soil moisture for plant growth varies with the stage of crop growth.
- ☆ Certain periods for crop growth and development are most sensitive to soil moisture stress as compared with other.
- ☆ The period is known as moisture sensitive periods.
- ☆ Critical period is used to show stages of growth when plants are most sensitive to shortage of water.
- ☆ Inadequate water supply during moisture sensitive periods will irrevocably reduce the yield and provision of adequate water and other management practices at other growth stages will not help in recovering the yield lost.
- ☆ In vegetables, transpire less water requires stress free moisture condition due to weak root system that is sparsely distributed and located in 15 to 20 cm layer of upper soil which gets dried up fast.
- ☆ Vegetable crops uses and transpire more water in later stages of growth during which moisture stress markedly reduces yield.
- ☆ All stages of growth is sensitive to soil moisture stress for crops where vegetative parts are of economic importance.
- ☆ Total growth and yield of perennial plants are sum effects of stress at each growth stage.
- ☆ Adequate water supply is required at flower bud initiation, flowering and fruit set.
- ☆ Flower bud formation, increases due to restricted water supply prior to flower bud initiation.
- ☆ The maximum benefit from scarce irrigation water is realized when irrigation is to be scheduled at moisture sensitive periods by withhold irrigations at other periods of lesser sensitivity.

5.3 NUMBER OF IRRIGATIONS

5.3.1 Appropriate irrigation system based on soil type

- ☆ It is seen that water is absorbed and moves slowly through clay soils, but once the soil gets wet, they retain significant amounts of moisture.
- ☆ Water is absorbed and moves quickly through sandy soils, but they retain very little which shows that water is applied quickly to clay soil has the tendency to run off rather than move in the soil.

- ☆ Irrigating of clay soils shows that water needs to be applied slowly over long period as a result of which, the site will not need irrigation for several days.
- ☆ Irrigation on sandy soils is applied quickly but for short periods.
- ☆ Irrigation times on sandy sites should be shorter, otherwise water moves beyond the root zone, and becomes unavailable to the plant and contributing to soil leaching.
- ☆ For efficient water use under certain weather conditions, sandy sites may need daily irrigation for short periods.
- ☆ Clay soils have greater capillary movement as compared to sandy soils.
- ☆ Quick water application on sandy soils will contribute to big wetting area which gives more soil volume for roots to exploit.

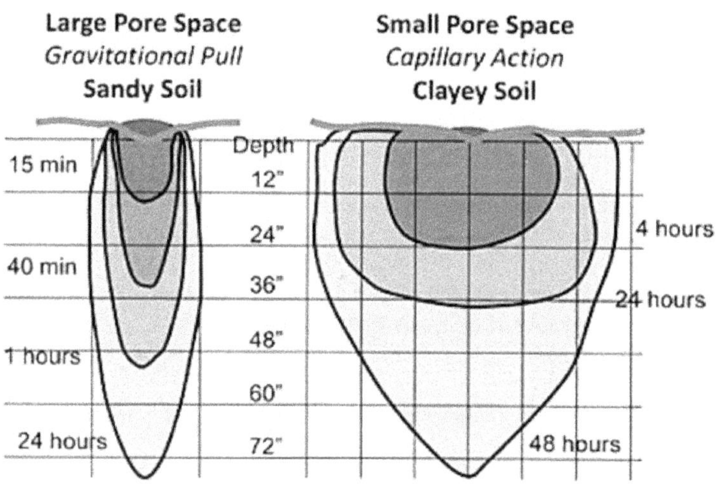

- ☆ Clay-based soils have small, flat, compact particles with large surface to volume ratios.
- ☆ Clay soil are often difficult to prepare for planting since they are slippery when wet and hard when dry, making timing for field operations critical to avoid damaging soil structure and getting proper soil tilth for planting.
- ☆ Sand-based soils are at the other end of the spectrum having comparatively large particles with small surface to volume ratios.
- ☆ Sand-based soil are easier to prepare for planting and can be worked shortly after significant rainfall.

Irrigation Technology as per soil

Basin irrigation

- ☆ It is the simplest of the surface irrigation methods.
- ☆ If the basins are small, they can be constructed by hand or animal traction.

☆ Their operation and maintenance is simple.

Furrow irrigation

☆ It requires accurate field grading which is mostly done by machines.

☆ The maintenance such as ploughing and furrowing are done by machines.

☆ It needs skill, organization and frequent use of foreign currency for fuel, equipment and spare parts.

Short level furrows

☆ It is called as furrow basins.

☆ It is like basins which can be constructed and maintained by hand.

Borders

☆ It requires highest level of sophistication.

☆ They are constructed and maintained by machines.

☆ The grading needs to be accurate.

☆ Machine operation requires a high level of skill, organization and usually foreign currency.

Soil type	Rooting depth of the crop	Net irrigation depth per application (mm)	Irrigation method
Sand	shallow	20-30	short furrows
	medium	30-40	medium furrows, short borders
	deep	40-50	long furrows, medium borders, small basins
Loam	shallow	30-40	medium furrows, short borders
	medium	40-50	long furrows, medium borders, small basins
	deep	50-60	long borders, medium basins
Clay	shallow	40-50	long furrows, medium borders, small basins
	medium	50-60	long borders, medium basins
	deep	60-70	large basins

5.3.2 Stages of Irrigation in Wheat Crop

☆ Required soil moisture is required for normal development of wheat plant at all the stage of growth.

- ☆ The crown root initiation stage and heading stage are critical stages when plant suffers most due to moisture stress.
- ☆ The schedule of irrigation needs to be followed for dwarf variety of wheat.
- ☆ In case of dwarf yielding variety as pre, sowing irrigation should be given and crop sown when the field become fit for the operation.

Schedule of irrigation for dwarf wheat

First Irrigation

- ☆ The first irrigation to standing crop be given 20-25 days after sowing at the crown root initiation stage.
- ☆ In cooler regions and in late sown wheat, it is desirable to apply first irrigation approximately 25-30 days after sowing.
- ☆ Delay in giving first irrigation will result in upsetting the synchronous tillering in wheat varieties, subnormal heads, poor root system and poor grain yield.
- ☆ It is most critical irrigation.

Second Irrigation

- ☆ The irrigation shows tillering stage which is 40-45 days after sowing.

Third Irrigation

- ☆ The irrigation shows late jointing stage which is within 70-75 days after sowing.

Fourth Irrigation

- ☆ The irrigation shows flowering stage which is within 90-95 days after sowing.
- ☆ Irrigation at this stage is important as during this period, plant suffer most from soil moisture deficiency.
- ☆ In this, the grain number and grain size are reduced considerably.

Fifth Irrigation:

- ☆ The irrigation shows dough stage which is within 110-115 days after sowing.
- ☆ In this, the total number of irrigations required varies as per soil type, winter rainfall amount of water applied per irrigation.

Irrigation schedule under limited supply of water

With limited supply of water, following irrigation schedule of irrigations should be adopted which strongly stresses importance of irrigation at CRI stage:

- ☆ Where only one irrigation is possible, give crown root initiation CRI stage 20-25 days after sowing.
- ☆ Where two irrigation are available, first irrigation be given at CRI stage and second flowering stage.
- ☆ Where three irrigation are possible, first irrigation be given at CRI stage while second irrigation at late jointing stage and third irrigation at milk stage.

It shows that each week delay in first irrigation from CRI stage results in yield reduction of 2-3 quintal per hectare.

5.4 IRRIGATION SYSTEM

- ☆ The irrigation system consists of:
- ☆ Intake structure

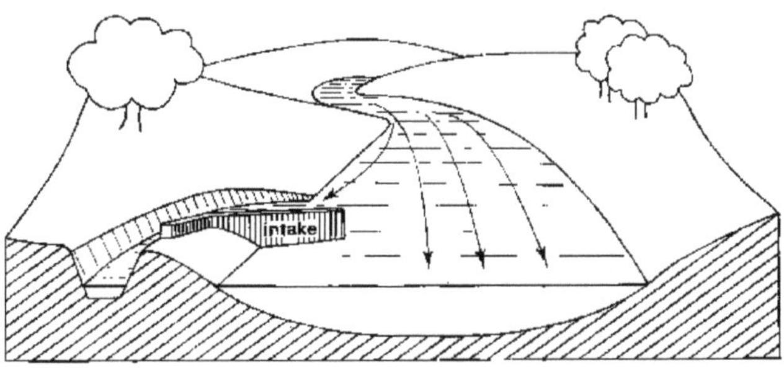

- ☆ The intake structure is built at the entry to the irrigation system.
- ☆ The purpose of intake structure is to direct water from the original source of supply (lake, river, reservoir etc.) into the irrigation system.

Pumping station

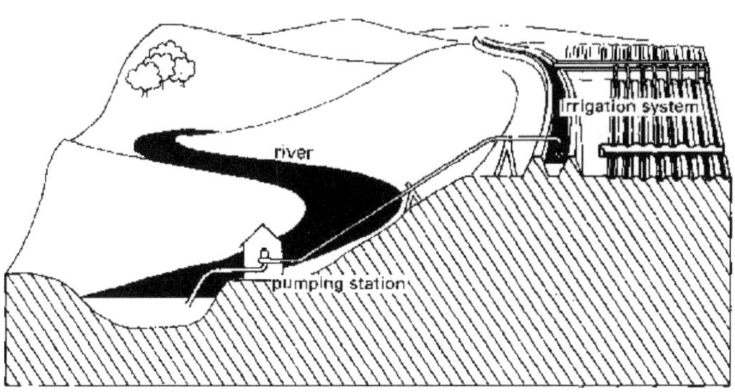

- ☆ Many times, irrigation water source lies below the level of the irrigated fields.

- ☆ The pump is used to supply water to the irrigation system.
- ☆ There are several types of pumps, but the most commonly used in irrigation is the centrifugal pump.

Conveyance system and distribution system

- ☆ The conveyance and distribution systems consist of canals transporting the water through the whole irrigation system.
- ☆ Canal structures are required for the control and measurement of the water flow.

Open canals

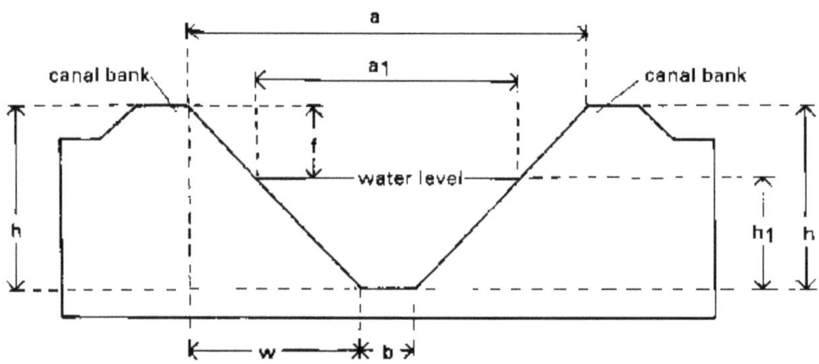

a = top width of the canal
a1 = top width of the water level
h = height of the canal
h1 = heigth or depth of the water in the canal
b = bottom width of the canal
h:w = side slope of the canal
f = free board (= h–h1)

- ☆ An open canal, channel, or ditch, is an open waterway whose purpose is to carry water from one place to another.
- ☆ Channels and canals refer to main waterways supplying water to one or more farms.
- ☆ Field ditches have smaller dimensions and convey water from the farm entrance to the irrigated fields.
- ☆ The most commonly used canal cross-section in irrigation and drainage, is the trapezoidal cross-section.

Field application system

- ☆ The traditional methods of applying water to the field is by bringing the water from the source of supply, such as a well, to each plant with a bucket or a water-can.
- ☆ Surface irrigation is application of water to the fields at ground level in which, either the entire field is flooded or the water is directed into furrows or borders.
- ☆ Furrows are narrow ditches dug on the field between the rows of crops.
- ☆ The water runs along the furrow as it moves down the slope of the field.

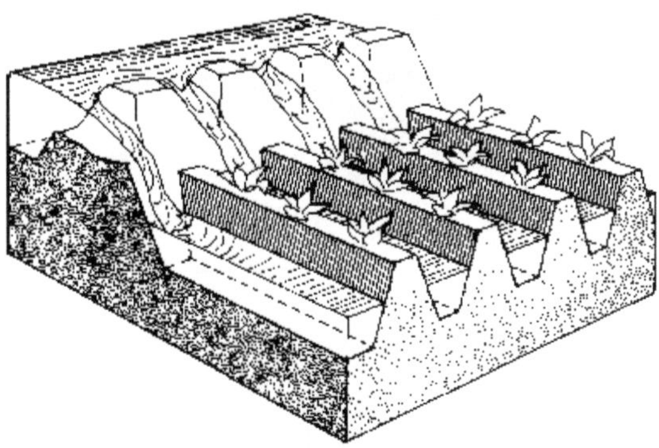

- ☆ The water flows from the field ditch into the furrows by opening up the bank or dyke of the ditch or by means of syphons or spiles.
- ☆ Siphons are small curved pipes that deliver water over the ditch bank.

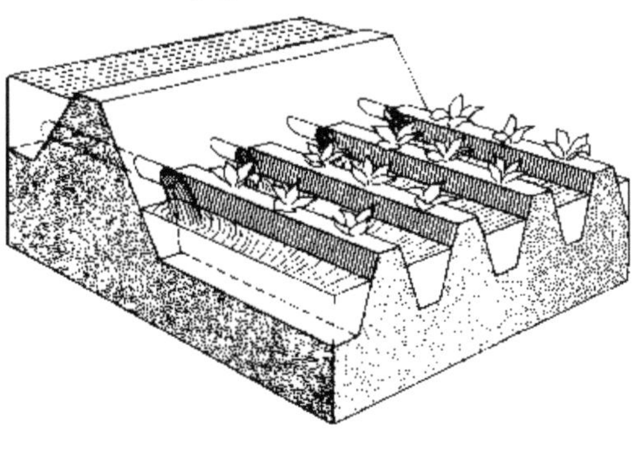

☆ Spiles are small pipes buried in the ditch bank.

☆ In border irrigation, the field to be irrigated is divided into strips by parallel dykes or border ridges.

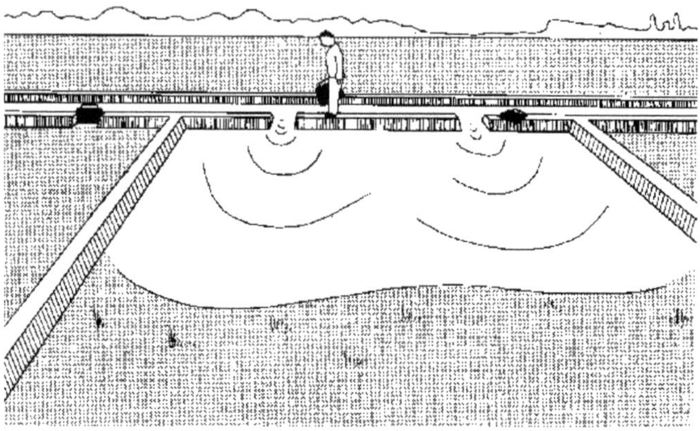

☆ In border irrigation, water is released from the field ditch on border through gate structures called outlets or by means of siphons or spiles.

☆ In basin irrigation, basins are horizontal, flat plots of land, surrounded by small dykes or bunds in which the banks prevent the water from flowing to the surrounding fields.

☆ Basin irrigation is commonly used for rice grown on flat lands or in terraces on hillsides.

- ☆ Trees can also be grown in basins, where one tree usually is located in the centre of a small basin.

Drainage system

- ☆ A drainage system is required to remove excess water from the irrigated land.
- ☆ The excess water may be e.g. waste water from irrigation or surface runoff from rainfall.
- ☆ Waste water can also be leakage or seepage water from the distribution system.
- ☆ Excess surface water is removed through shallow open drains.
- ☆ Excess groundwater is removed through deep open drains or underground pipes

5.5 MICRO IRRIGATION SYSTEM

5.5.1 Introduction

- ☆ It is the slow application of water as discrete or continuous drips, tiny streams or miniature spray system on, above, or below the soil by surface drip.
- ☆ It is applied through emitters connected to water delivery line by low-pressure delivery.
- ☆ It has very small openings known as emitters, that are designed to discharge the water at a known rate for operating pressure.
- ☆ It needs more planning than overhead sprinkler systems as the emitter systems deliver water directly to the root system.
- ☆ It encompass several ways of water application to the crop such as:
  - drip irrigation
  - spray irrigation
  - subsurface irrigation
  - bubbler irrigation

5.5.2 Types of Micro irrigation systems

1. Drip Irrigation

Components

- ☆ Pump station
- ☆ By-pass assembly
- ☆ Control valves

- ☆ Filtration system
- ☆ Fertilizer tank /Venturi
- ☆ Pressure gauge
- ☆ Mains / Sub-mains
- ☆ Laterals
- ☆ Emitting devices
- ☆ Micro tubes

- ☆ It is also known as trickle irrigation system.
- ☆ It is frequent, slow application of water to soils by using mechanical devices known as emitters or applicators which are located at selected points along the delivery lines.
- ☆ The emitters dissipate the pressure from the distribution system using orifices, vortexes and tortuous or long flow paths which allow limited volume of water to discharge.
- ☆ The emitters are placed on the ground or are buried.
- ☆ In this, emitted water moves in the soil system mainly by unsaturated flow.
- ☆ The wetted soil area for widely spaced emitters is normally elliptical in shape.
- ☆ In this, the water is led to the field by pipe system where next to row of plants or trees, a tube is installed.
- ☆ At regular intervals, near the plants or trees, a hole is made in the tube and equipped with an emitter.
- ☆ The water is supplied slowly, drop by drop, to the plants through these emitters

Advantages

- ☆ Increases production & productivity.
- ☆ Improves quality and ensure early maturity of crops.
- ☆ Water Saving up to 40% - 70%.

- ☆ Controls weed growth, saving of fertilizer (30%) and labour cost (10%).
- ☆ Fertigation / Chemigation can be made efficiently.
- ☆ Control diseases.
- ☆ Use of saline water is possible.
- ☆ Soil erosion is eliminated.
- ☆ Suitable for uneven / undulating land.
- ☆ High Water Use Efficiency

2. Sprinkler irrigation

- ☆ It creates artificial rainfall is created where the water is led to the field by pipe system where water is under pressure.
- ☆ In this, the spraying is done by using many rotating sprinkler heads or spray nozzles or single gun type sprinkler.
- ☆ In this, the sprinklers deliver the water to the plant and/ or soil which takes water from the source and distribute the water uniformly over an area in droplet form.
- ☆ To cover larger area, sprinkler needs to throw water to a considerable distance.
- ☆ If the sprinkler is good, it will take many factors into account which includes water supply, soil, crop, topography, and atmospheric conditions.

Spray-type sprinklers

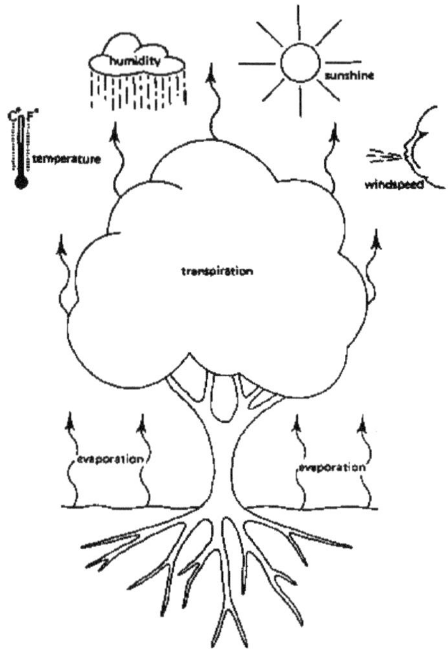

- ☆ The spray nozzles require considerably less pressure and so energy than impact sprinklers.
- ☆ Spray nozzles are installed at the end of drop tubes/hoses to release water closer to the crop canopy to reduce wind and evaporative losses.
- ☆ The spray nozzle have smaller wetted diameter that causes very high application rates at the outer end of lateral.
- ☆ In heavy soil, it leads to application rates that exceed the soil infiltration rate which results in runoff.

5.6 QUANTITY OF WATER REQUIRED ACCORDING TO STAGES OF GROWTH

- ☆ The plant roots suck or extract water from the soil to live and grow.
- ☆ The main part of the water does not remain in the plant, but escapes to the atmosphere as vapour through the plant's leaves and stem which is known as transpiration.
- ☆ Transpiration happens mainly during the day time, where water from an open water surface escapes as vapour to the atmosphere during the day which same happens to water on the soil surface and to water on leaves and stem of a plant.
- ☆ The water need of a crop comprises of transpiration and evaporation known as evapotranspiration.

- ☆ The water need of a crop is usually expressed in mm/day, mm/month or mm/season.
- ☆ A certain crop grown in a sunny and hot climate needs per day more water than the same crop grown in a cloudy and cooler climate.
- ☆ Apart from sunshine and temperature, other climatic factors which influence the crop water need includes humidity and windspeed.
- ☆ When it is dry, the crop water needs are higher than when it is humid.
- ☆ In windy climates the crops will use more water than in calm climates.

5.6.1 Effect of climatic factors on water need of crop

Climatic Factor	Crop water need	
	High	Low
Sunshine	sunny (no clouds)	cloudy (no sun)
Temperature	hot	cool
Humidity	low (dry)	high (humid)
Windspeed	windy	little wind

- ☆ The highest crop water needs are found in areas which are hot, dry, windy and sunny.
- ☆ The lowest values are found when it is cool, humid and cloudy with little or no wind.
- ☆ Single crop grown in different climatic zones will have different water needs.
- ☆ For calculating water need of crop, it is useful to take a certain standard crop or reference crop and determine how much water is required by the crop per day in various climatic regions.

5.6.2 Growth Stage of Crop as per water required

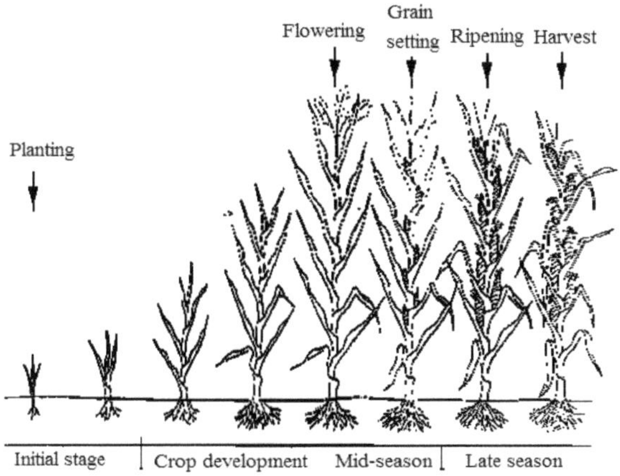

- ☆ A fully grown crop will need more water than the crop which has just been planted.
- ☆ The crop water need or crop evapotranspiration consists of transpiration by the plant and evaporation from the soil and plant surface.
- ☆ If crop are very small, the evaporation will be more important than the transpiration.
- ☆ If the plants are fully grown the transpiration is more important than the evaporation.
- ☆ At planting and during the initial stage, evaporation is important than transpiration and the evapotranspiration or crop water need during initial stage is estimated at 50% of crop water required during the mid - season stage, when the crop is fully developed.
- ☆ At crop development stage, crop water requirement gradually increases from 50% of maximum crop water to maximum crop water need.
- ☆ The maximum crop water need is reached at the end of the crop development stage which is the beginning of the mid-season stage.
- ☆ At late season stage, the period during which the crop ripens and is harvested, a distinction can be made between fresh harvested crops and dry harvested crops
- ☆ With fresh harvested crops, water requirement remains the same during the late season stage as it was during the mid-season stage.
- ☆ With dry harvested crops at late season stage, these crops are dried out and even die, so their water needs during the late season stage is minimal.

5.7 OPTIMUM WATER LEVEL TO BE MAINTAINED TO AVOID EXCESSIVE WATERING AND SEEPAGE LOSSES

5.7.1 Optimum water level

- ☆ Different crops need different type of soils, amounts of nutrients and different types and amounts of water.
- ☆ The amount of water required by the plant is dependent on the growing season and the climate where it is grown.
- ☆ Selecting right crop for given soil conditions and climate will save water requirements for irrigation.
- ☆ The water need of a crop is expressed in mm/day, mm/month or mm/season.

☆ If the water need of certain crop in very hot, dry climate is 10 mm/day, then each day the crop needs a water layer of 10 mm over the whole area on which the crop is grown.

☆ The crop water need is the depth of water required to meet the water loss by evapotranspiration.

☆ The crop water need refers to crop grown under optimal conditions, i.e. a uniform crop, actively growing, completely shading the ground, free of diseases, and having favourable soil conditions.

5.7.2 Correct water requirement for wheat crop

☆ The loss of water by evaporation and transpiration from wheat field is evapotranspiration which directly influences the growth and yield of the crop.

☆ For harvesting good crop of wheat, about 40 cm depth of water is used as evapotranspiration.

☆ There is a need for 4-5 irrigations for wheat.

☆ In sandy loam soil 6-8 irrigations may be required whereas in heavy clay soil 3-4 irrigations are considered to be sufficient.

Ample/Shortage of water availability

☆ If the water available is adequate then irrigation should be given at all the stages.

☆ If water is in short supply then irrigations should be given at certain critical stages.

☆ Maximum reduction takes place if the irrigation at the critical growth stage is missed.

☆ If the water is available for less than adequate number (5-6) of irrigations then crop should be irrigated as:

- For one irrigation: it should be applied at crown root initiation in 20-25 days after sowing.
- For two irrigations: first irrigation at crown root initiation and second at flowering stage.
- For three irrigations: first irrigation at crown root initiation, second at late jointing and third at milk stage.
- For four irrigations: crop to be irrigated at crown root initiation, late jointing, heading and milk stage.

5.7.3 Problems related to over/under watering

The potential drawbacks of over/under watering include:

Under-watering

- ☆ Loss in market value through yield reduction
- ☆ Reduction in fruit size and quality

Over-watering

- ☆ Unwanted vegetative growth
- ☆ Losses of valuable water to the water table
- ☆ Irrigation water travel over soil cause erosion where excessive displacement of top soil lead to soil fertility, clog drainage ditches and streams, harm aquatic habitats, increases the need for water treatments.
- ☆ Irrigation cause pesticides, pathogens and weeds to spread during irrigation
- ☆ Causes runoff
- ☆ Increased operational costs
- ☆ Leaching of nutrients lead to algal growth, salinity an nitrate build ups elsewhere in the catchment
- ☆ Downgraded product quality and reduced yield.
- ☆ Higher operational costs for producer
- ☆ Pressure on water resources with increasing demand for water use by urban dwellers

5.7.4 Precautions of over irrigation of crop

- ☆ To avoid crop stress, do not allow soil water to fall below 50% of plant available water capacity.
- ☆ Make sure water is available for 2 to 3 days before the crop reaches its refill point.
- ☆ The period leading up to and including flowering is the most sensitive to water stress.
- ☆ Stress at this time will reduce the number of heads per plant, head length, and number of grains per head.
- ☆ The application of a fixed water depth at each irrigation can lead to deep drainage losses.
- ☆ It is not necessary to refill the soil profile at each irrigation.

- ☆ Overhead systems are especially suited to application of small irrigation depths, but application depths can also be reduced with surface irrigation systems by increasing siphon flow rates and reducing irrigation runtimes.
- ☆ Crops can only extract water from their effective root zone, so depth of soil wetted by irrigation needs to be adjusted during the season to respond to increases in root zone depth and irrigation wetted front should not go deeper than the effective crop root zone.

Chapter 6

INTEGRATED PROTECTION MANAGEMENT

6.1 VARIOUS BIOTIC FACTORS THAT AFFECT THE WHEAT CROP SUCH AS INSECTS, DISEASES AND NEMATODES

6.1.1 Biotic factors

- ☆ Biotic factors such as pests, insects and diseases reduce the crop production.
- ☆ A pest causes damage to our crops by feeding.
- ☆ Weeds also reduce crop productivity by competing with the main crop for nutrients and light.
- ☆ The pests, weeds and vertebrate pests like rodents are main causes for crop yield reduction.
- ☆ The crop lost its tune of 25% due to insects and diseases which includes nematodes up to 20%, vertebrate pests including rodents about 6-8% etc.

6.1.2 Affect on wheat crop by insects

Aphids

- ☆ It is available in all wheat grown areas in NWPZ and Peninsular India.
- ☆ It sucks the life out of fruitful wheat harvest and lives in wheat to suck the sap from the plant.

- ☆ They exist in different stages such as winged (alates), wingless (apterous) sexual and asexual forms.
- ☆ It damages the plant's livelihood through sustenance and also passes barley yellow dwarf virus to the plant.
- ☆ They rapidly spread through asexual reproduction where females give rise directly to nymphs rather than eggs.
- ☆ In this, infestation takes place during second fortnight of January till crop maturity.
- ☆ If feeds in sufficient numbers, they cause considerable damage, but under normal conditions, losses are not much.
- ☆ To control the pest in wheat crop, chemical pesticides are recommended if they crosses 10 aphids per tiller during vegetative phase and 5 aphids per tiller during reproductive phase.
- ☆ The spray of imidacloprid @ 20 g a.i. per ha is applied on border rows and if infestation is severe then it should be used in entire field.
- ☆ The natural enemies present in the field helps in controlling the pest population.

Brown Wheat Mite

- ☆ It is found in most of the wheat growing areas and in rainfed conditions.
- ☆ It is a pest which is found in humid and warm conditions of irrigated areas.
- ☆ The wheat cause damage through mite infestation.
- ☆ If present in large number, they cause silvery flecking on leaves.

- ☆ Each mite is too small to be visible with naked eye without effort and is seen by shaking infested leaves.
- ☆ They do not cause any production constraint in wheat, hence no management practices are required for it, but it can change cropping sequence in future.

Army Worm

- ☆ It is found in warmer climates in central and northern plains.
- ☆ In this, larvae are found in cracks of soil which hides during the day and feeds during the night or early morning.
- ☆ These are among the most common and damaging insects to wheat.
- ☆ They don't infect the plant as aphids do, but feed sufficient wheat during their lifetimes.
- ☆ It eats 34 inches of plant crop in just three to five days causing the plant in serious damage.
- ☆ In wet and humid weather, they feed during day time.
- ☆ They survive during summer on subsequent crops and continue to exist in rice crop before wheat crop comes in field.
- ☆ It is seen in rice-Wheat rotation and in areas of rice stubbles/straw which remain in the fields.

Legume Pod-borer

- ☆ It is a polyphagous insect which attacks on many legumes as pod border.
- ☆ It damages wheat ear heads at grain development stage when major hosts are not present.

☆ The damage which causes due to it is below the economic threshold level.

☆ It is found mainly in northern and central parts of India.

☆ Wheat serves as bridge host for carrying of such polyphagous pest.

Termites

☆ It is found mainly in areas of northern and central India and also in some portions of peninsular India.

☆ It attacks the crop at various growth stages, from seedlings to maturity.

☆ The damaged cause to the crop can be easily uprooted and look wilted and dried and damaging roots shows yellowish effect.

☆ To protect seed and to treated soil for the crop, chemicals such as endosulfan, chlorpyriphos and carbosulfan are used.

Stink bugs

☆ The bug's saliva is toxic to the wheat plant.

☆ The bug on creating a single puncture on wheat stalk, could kill the entire plant.

☆ It eats the wheat kernels and head of the wheat and cause crop to wither and die.

☆ To control the damage, pesticides such as fenpropathrinm, dimethoate, endosulfan and formetanate hydrochloride are to be used.

6.1.3 Wheat Diseases

Powdery mildew

☆ It is a white, powdery patches which is formed on upper surface of leaves and stem.

☆ It is a form of grayish white powdery growth that is present on the leaf, sheath, stem and floral parts.

☆ The powdery growth later on turns to black lesion and causes drying of leaves and other parts.

- It is a fungus which remains in high hills during summers in infected plant debris as dormant mycelium and asci.
- It is mainly spread by asciospores and through airborne conidia.
- It infects the plants during the period of high humidity and cool to moderate temperatures.

Loose smut

- It is a seed borne disease which results in infection during flowering by wind-borne spores.
- In this, the infection remains dormant inside the seed but the plants growth bears with infected inflorescence.
- In this, entire inflorescence commonly affects and appears as a mass of olive-black spores, which is covered by thin gray membrane.
- On rupturing of membrane, the head appears powdery.
- It causes infection by sowing infected seeds.
- Infection is favored by cool, humid conditions during flowering period of host plant.

Brown rust

- It is present on the upper leaf blades and also on sheaths, glumes and awns.
- In this, pustules are circular or elliptical and are smaller than stem rust which are not coalesce, but contain masses of orange to orange-brown Urediospores.

- ☆ The pathogen is found in summers in low and mid altitudes of Himalayas and Nilgiris.
- ☆ In this, the infection is started from wind deposits urediospores in eastern Indo-gangetic plains in mid of January where it multiplies and moves westwards by March

Stripe rust /Yellow rust

- ☆ It mainly takes place on the leaves and than on leaf sheaths and stem.
- ☆ It shows presence of bright yellow pustules which appear on leaves at early stage of crop arranged in linear rows as stripes.
- ☆ In this, stripes are of yellow to orange yellow in colour with teliospores which are dull black in colour.
- ☆ The pustules of stripe rust with yellow to orange-yellow urediospores forms narrow stripes on the leaves.
- ☆ The pustules is also found on leaf sheaths, necks and glumes.
- ☆ In this, inoculum survives in form of uredospores /teliospores in northern hills at off season on self sown crop or volunteer hosts that shows good source of inoculums.
- ☆ The disease is high when temperature is 10-20°C and very humidity.

Black rust

- ☆ It is found on almost all aerial parts of wheat plant and are common on stem, leaf sheaths, upper and lower leaf surfaces.
- ☆ The pustules are dark reddish brown which occurs on both sides of leaf, stems, and on spikes.

- ☆ Before spore masses break by epidermis, the infection sites feel rough and after breaking of spore masses, the surface tissues take on ragged and torn appearance.
- ☆ It mainly spread through uredospores from southern hills
- ☆ The disease is common in moisture and temperature above 20°C.

Flag smut

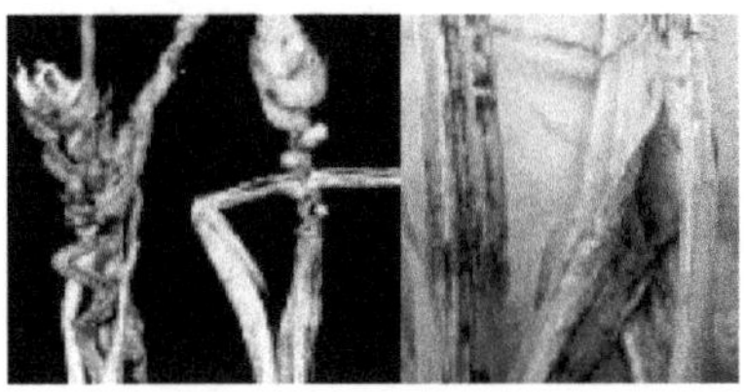

- ☆ Its symptoms is seen on stem, clum and leaves from late seedling stage to maturity.
- ☆ The seedling infection lead to twisting and drooping of leaves followed by withering.
- ☆ Grey to grayish black sori takes place on leaf blade and sheath which carries black powdery mass of spores.
- ☆ It is a seed and soil borne disease.
- ☆ The smut spores are viable for more than 10 years.
- ☆ The main infection occurs by sowing infected seeds or by resting spores present in the soil.
- ☆ The disease is common in temperature of around 18-24°C and also when relative humidity is 65% or above.

Hill bunt/Stinking smut

- ☆ It is a fungus which attacks the seedling of 8-10 days old and became systemic and grows along the tip of shoot.
- ☆ During flowering hyphae concentration in inflorescence and spikelets it transforms ovary in smut sorus of dark green color with masses of chlamydospores.

- ☆ When the diseased plants mature early, it will affect all the spikelets.
- ☆ In this, pathogen survives in seeds and sows the seeds which serves as a source of primary infection.
- ☆ The disease is common in temperature of around 18-20°C and also in high soil moisture.

Karnal bunt

- ☆ It symptoms are difficult to distinguish in field as incidence of infected kernels on a given head is low.
- ☆ It results as spreading of glumes due to sorus production and is not extensive as that of common bunt.
- ☆ The symptoms of it are most readily detected on seed after harvesting.
- ☆ It is a seed borne disease and sowing of infected seeds serves as main source of infection.
- ☆ The disease is common in temperature of 18-20°C and also in high soil moisture.

Leaf blight

- ☆ It is a reddish brown oval spots that appear on young seedlings with bright yellow margin.
- ☆ In severe cases, many spots coalesce which causes leaves to dry.
- ☆ It is a complex disease which mainly spread by externally seed-borne and soil borne conidia.
- ☆ The disease is common in temperature of 25°C and high relative humidity.

Foot rot

- ☆ The disease mainly occurs in seedlings and roots which turns rootlets to brown.
- ☆ It lead to pale green seedling with stunted growth.
- ☆ It is a fungus which produces sporangia and zoospores and oospores.
- ☆ The disease is soil borne, where pathogens survives in soil.
- ☆ Its mainly spread through soil and irrigation water.
- ☆ The disease is common in wet weather and high rainfall areas.

Head scab/Fusarium leaf blotch

Leaf blotch:

- ☆ The blotching caused by the organism is evident on leaves at about late joint to early boot growth stage.
- ☆ The young lesions occur as oval to elliptical, grayish green mottled areas, which is located where the leaf bends.
- ☆ The lesions enlarge rapidly and develops into large eyespot blotches with bleached or light grey centers.
- ☆ In this, the leaves tend to split or shred, beginning at the centers of the lesions.

Head Scab:

- ☆ It is caused by fungus.
- ☆ Its symptoms include tan or light brown lesions with one or more spikelets.
- ☆ Some diseased spikelets have dark brown discoloration at the base and an orange fungal mass along lower portion of glume.
- ☆ In this, grain from plants by Fusarium head blight is shriveled with white chalky appearance.
- ☆ Infected florets becomes slightly darkened and oily in appearance.
- ☆ The disease is soil borne and inoculums of fungi survive in soil.
- ☆ Spores are produced on crop debris left on or near the soil surface.
- ☆ The spores are transmitted to leaves by the wind or by splashing rain.
- ☆ The disease is common in cool, moist weather.

Helminthosporium leaf blotch

- ☆ Lesions caused by the disease is oval in shape and are dark brown in color.
- ☆ On maturing of lesions, centers turn light brown to tan color surrounded by irregular dark brown ring.
- ☆ It mainly affects on the lower leaves, starting as chlorotic flecks or spots.

- ☆ The infection sites enlarge, turn dark brown and coalesce.
- ☆ When the disease is severe, it affects the leaves or leaf sheaths and makes them to die prematurely.
- ☆ It is a seed and soil borne diseases in which the inoculums is present in seeds and soil as main source of infection.
- ☆ The disease is common in very humid and higher rainfall areas.

Seedling blight

- ☆ The symptom of the disease is seen when ears become infected during early flowering stages.
- ☆ Later infections result in infection of grain but without obvious bleaching of ears.
- ☆ Fusarium lesions started in leaf sheath at stem base where crown roots split the leaf sheath when emerging.
- ☆ The infection spreads in the leaf sheath which causes long dark brown streaks at stem base.
- ☆ The ear blight phase of disease causes yield loss and result in mycotoxin production in grains.
- ☆ An important source of Fusarium for wheat crops is seed, but fungus can also survive on debris in the soil.
- ☆ The disease is common in relatively high soil moisture and in high soil temperature.

6.2 STAGES OF WHEAT CROP GROWTH WHEN IT IS VULNERABLE

- ☆ The wheat crop will soon be in heading stage, if wheat becomes most susceptible to Fusarium head blight (FHB) or head scab. There are many factors which makes the wheat crop in development of Fusarium head blight:

Previous crop:

- ☆ The fungus that causes head scab survives in residues of many grass crops.

- ☆ The fungus is a pathogen of corn and most severe disease often occurs when wheat is planted in fields with large amounts of corn residue left on the soil surface.

Variety susceptibility:

- ☆ Most wheat varieties are susceptible to head scab, but some varieties are especially vulnerable to the disease.

Weather conditions:

- ☆ FHB infection of wheat takes place at flowering or during the early stages of the grain filling period.
- ☆ The time period clearly influences the amount of disease present while weeks preceding flowering are also important.
- ☆ The reproduction of fungus that causes head scab is favored by frequent rainfall and extended periods of high relative humidity.
- ☆ The worst epidemics of scab takes place when conditions are favorable for reproduction of fungus prior to flowering followed by few days that are conducive for infection during flowering or early stages of grain fill.

Recommendations

- ☆ If above factors show moderate to high risk of head scab, a fungicide is required to help suppress the disease, whose selection is important.
- ☆ The fungicides in strobilurin class offer little protection and are not recommended for control of head scab.
- ☆ The triazole fungicides Folicur, Prosaro, Proline, and Caramba are available for use and can help suppress disease development by 40 to 60 percent.
- ☆ The fungicides should be applied to heads to have any activity against scab, and should be applied as close to flowering as possible.
- ☆ The growth stage cut-off for these fungicides is 50% flowering, having 30-day pre-harvest interval.

6.3 Symptoms caused by insects, diseases and nematodes

6.3.1 Symptoms caused by Nematodes

- ☆ Nematodes are microscopic unsegmented roundworms that are one of the most numerous life forms on earth.
- ☆ Plant parasitic nematodes live in plant roots and other plant parts causing disease.

- ☆ Root-knot nematode; Symptoms
- ☆ Infect the plants in patches in the field
- ☆ Forms galls on host root system
- ☆ Roots branch profusely starting from gall tissue which lead to beard root
- ☆ Infected roots become knobby and knotty
- ☆ In infected plants, root system is reduced and rootlets are absent.
- ☆ It hampered plant roots in their function
- ☆ Plants wilt during the hot part of day, under dry conditions
- ☆ It predisposes plants to fungal and bacterial root pathogens

Molya nematode; symptoms

- ☆ It attack the plants and make them stunted.
- ☆ It gives shriveled unhealthy appearance.
- ☆ It stimulates the formation of branched rootlets.
- ☆ In this, main root remains short or bunchy, bearing small galls.
- ☆ In severe infestation, seedlings fails to come out of soil.
- ☆ Plants which escapes early damage results in short stalks and ears, giving poor harvest.

Wheat-gall nematode; symptoms

- ☆ It creates basal swelling of stem.
- ☆ It makes crinkling, twisting and diverge of leaves.
- ☆ It makes stunted growth and increases tillering
- ☆ It lead to diverge nature of spikes.
- ☆ It reduce ear head.

Seed gall nematode; Symptoms

- ☆ It distorts leaves and stems prior to heading.
- ☆ During maturity of diseased plants, galls are formed in the florets, which replace kernels.
- ☆ Seed galls in cereals are ear cockle which are similar in shape of seeds they replace and are dark brown in color.
- ☆ In this, heavy motile larvae are present in galls which become active after galls have been moistened.

6.3.2 Symptoms caused by insects

- Many insects are associated with wheat.
- Some insects uses plant for food and damage it to such an extent that it cannot produce a good quality yield.
- Others insects co-exist, by plant for food but not causes economic damage.
- Not all insects found in wheat are detrimental; many natural enemies of pest species are attracted to wheat fields by presence of pest insects which they uses as food and hosts.

Insects pests of seeds

Greater False Wireworm

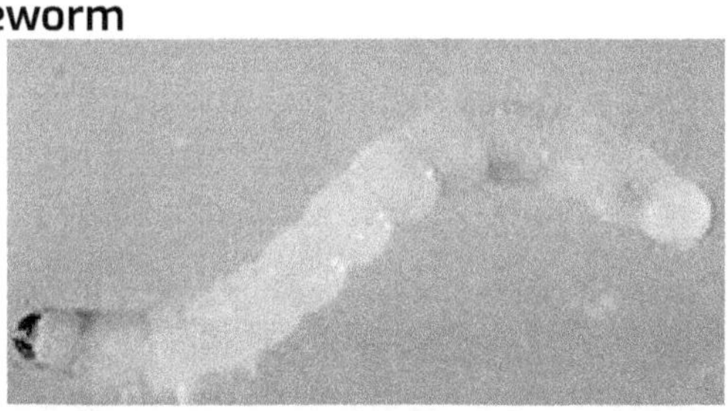

- It is the larval stage of dark-colored beetles with long legs seen on soil surface and under plant litter.
- In this, larvae feed on the seed and stems of seedlings below soil surface which causes stem to have tattered appearance.
- The adult beetles damages emerging seedlings which eventually wilt and die.

Lesser False Wireworm

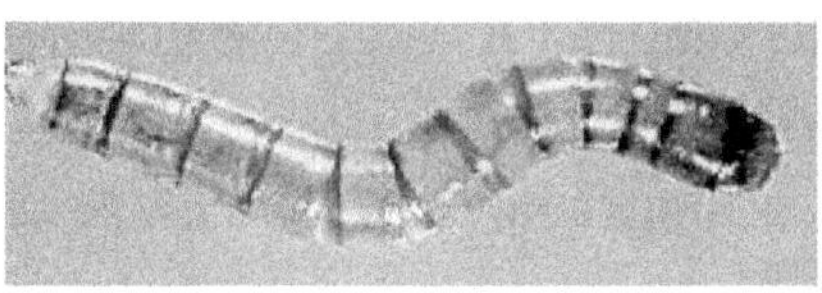

- It occurs sporadically throughout summer rainfall region.
- It is the larval stage of dark coloured beetle 'Dusty Surface Beetle'.
- The larvae feeds on seed and stems of seedlings below the soil surface which causes stem with tattered appearance.
- It makes the plants wilt and die leading to seedling density-loss in field.

Black Maize Beetle

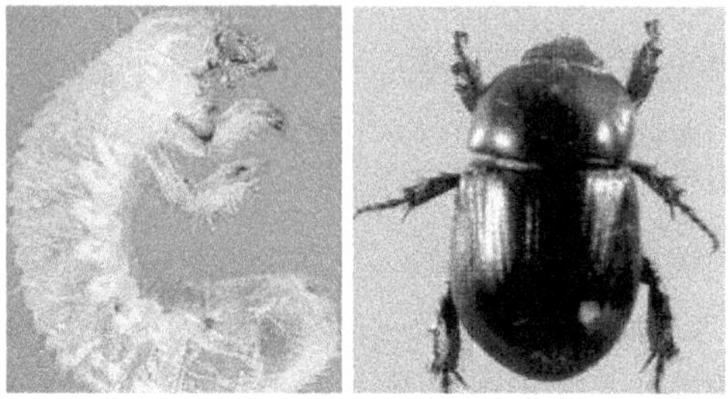

- It occurs sporadically throughout the summer rainfall region.
- The adult beetles have shiny black colour and are 12 - 15 mm in length.
- They are found in or on the soil surface layer.
- It feeds on the seedling stems at or just below the soil surface.
- It lead to big holes with ragged edges after feeding which results in plant death and seedling density-loss.
- Its larvae feed on decaying organic material and do not damage the crop.

Western Province Grain Worm

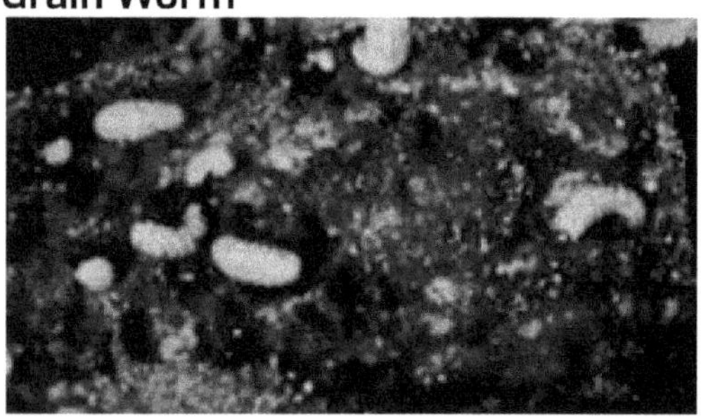

- It is a sporadic and unpredictable pest of wheat under intense rainfall conditions.
- In this, C-shaped larvae are legless and can reach 10 mm in length.
- It exist in white to cream coloured with small shiny brown heads.
- Its larvae feed on the stems of seedlings below the soil surface which causes stem to have tattered appearance.
- It makes the plant to wilt and die leading to poor plant standing in fields.
- The adult beetles feed on the leaves and will not economically damage the plant.

Sandveld Grain Worm

- It is 10 - 12 mm in length.
- The adult beetles are blue-green to metallic green and is 6 mm long.
- In this, the pupae are present in cells below the soil surface.

- ☆ Its larva is of cream colour with dark brown head is feed on stem of plant below the soil surface which lead small plants to die.
- ☆ Big plants survive with their feeding damage and tillering of plants gets reduced.

Insect pests of leaves and stems

Russian wheat Aphid

- ☆ It is a small spindle shaped pale yellow-green to grey-green aphid with short antennae and double tail.
- ☆ The wheat plants infected with RWA become prostrate which makes the leaves to roll out tightly.
- ☆ It leads to longitudinal, white or pale yellow stripes on leaves, which turns purple on cold conditions.
- ☆ The tightly rolled leaves trap the emerging ears.
- ☆ It yield loss to more than 60%, if not controlled chemically.
- ☆ It is more prone to damage wheat crop during the period between emergence of flag leaf and awn.

Greenbug

- ☆ It occurs sporadically in summer and winter rainfall regions and causes more problem in summer rainfall areas.
- ☆ The adults are of the size 1.8-2 mm and are pear shaped.
- ☆ Its colour is usually light yellowish green with dark green stripe on the back.
- ☆ Its cornicles are of pale green colour with distinct black tip.
- ☆ It feeds on upper side of leaves and turns the leaf to yellow which later on turns to brown chlorotic marks.

Oat Aphid

- ☆ It occurs annually in irrigated areas, but sporadically in summer rainfall region.
- ☆ Its body is rounded with adults having length from 1.5-2 mm.
- ☆ Its antenna is short with dark green body colour having red brown area visible.
- ☆ Its cornicles are short and are clearly visible.
- ☆ It feed on the leaves and stems of wheat plant.
- ☆ They migrate into ears during the latter part of the season where they extract sap directly from developing kernels.
- ☆ Its symptoms are visible only during high aphid population pressure which gives yellow plants.

Brown Ear Aphid or English Grain Aphid

- ☆ It occurs in wheat irrigation areas but sporadically in summer rainfall areas.
- ☆ Its length is around 2-3 mm.
- ☆ It is brown or green in colour.
- ☆ It feeds on the leaves of plants but after ear emergence, moves towards the ears.
- ☆ Its damage is not defined as when present in high numbers may turn leaves into yellow colour.

Rose Grain Aphid

- ☆ It occurs mainly on wheat produced under irrigation on dry land in winter rainfall regions.
- ☆ It measures 1.5-3 mm in length.
- ☆ It is light to yellowish green in colour with dark green longitudinal stripe on the back.
- ☆ Its cornicles are long without any conspicuous colour.
- ☆ It is found in small colonies on underside of leaves.
- ☆ It feeds on the stems and turn leaves to yellow during high populations of feeding on plants.

Grain chinch bug

- ☆ It is a sporadic pest of wheat restricted to small grain production areas.
- ☆ It is black in colour and measures 4-5 mm in length.
- ☆ It has lighter markings on its wings and emits vile odour.
- ☆ It is a sap sucking insect which spreads high infestations by making plant yellowish in colour.
- ☆ Its symptoms increases by warm and under dry conditions.

Bagrada Bug

- ☆ It occurs sporadically in small grain production areas of summer rainfall region.
- ☆ The adult is of shield shaped which is 4-7 mm in length.
- ☆ It is black in colour with orange and yellow markings.
- ☆ It feed on the leaves and stems but inflict damage during high populations.
- ☆ Its feeding symptoms is characterised by white rosette shaped markings on leaf surface.

Leafhopper

- ☆ It occurs sporadically in wheat produced irrigated areas.

- ☆ It has a soft body which is around 2.5 mm in length.
- ☆ Its head is yellow with two black spots on either side with full brown body.
- ☆ It transmit maize streak virus to wheat, after feeding on sap of infected maize or grass species.
- ☆ The immature virus infected plants have rosette or stunted appearance with longitudinal yellow stripes on the leaves.
- ☆ It leads to bent or curled leaves, when plants are infected.
- ☆ Direct feeding damage inflicts to young plants during high populations.

Brown Wheat Mite

- ☆ It is a sporadic pest in summer rainfall regions and in drought conditions.
- ☆ It feeds on sap from leaves by inserting two needle like stylets on leaf which causes dapple appearance on leaf surface.
- ☆ During high population, leaves result in bronze appearance while some leaves die due to intense feeding.
- ☆ It makes the plants stunt, head poorly, and turns it to white.
- ☆ The damage takes place under drought conditions.

Black Sand Mite

- ☆ It is of small grains and pastures which are black in colour.
- ☆ Its feeding result on older leaves, shows silvery white scars adjacent to main nerve.
- ☆ During high population, leaves may get discolour and plant gets wrinkled appearance which makes small plants to die.

Grain Slug

- ☆ It occurs throughout the small grain production areas.
- ☆ Its larvae feed on the upper side of the leaf between the veins resulting in white longitudinal stripes on the leaves.

6.3.3 Symptoms caused by diseases

Bacterial leaf streak and black chaff

- It is caused by sudden appearance of water-soaking on leaves.
- It is light brown and has elongated lesions on upper leaves.
- The lesions quickly dried out and turn into necrotic streaks on leaves.
- The black stripes occur on glumes and purple black lesions appear on rachis and peduncle in case of head infection.
- It is spread through infected seed and splashing water.

Basal glume rot

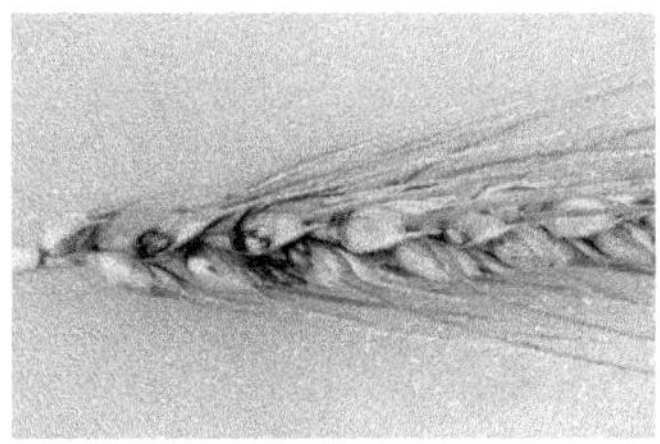

- It causes dull brown to black discoloration of glumes, pronounced on inner side.
- The seeds are shriveled, if infection is severe which makes discolored glume.
- It forms small water soaked lesions on leaves.
- It is spread mainly by infected seed.

Barley yellow dwarf

- It shows yellowing leaves mainly flag leaves.
- The stunted plants result from shortened internodes.

- The leaves are red, purple, orange, green or brown and are distorted.
- It is caused by virus.
- It is transmitted by few species of aphid.

Common bunt

- ☆ It makes slender heads which takes long time to turn color than healthy heads.
- ☆ The glumes spread apart to reveal spori or bunt balls that are similar size to normal kernel.
- ☆ It breaks the bunt balls on harvest and give off a fishy odour.
- ☆ It is caused by virus.
- ☆ It is spread through infected seed although spores are spread by wind.

Ergot disease

- ☆ It causes grains in the head which are replaced by dark purple to black sclerotia.
- ☆ The ergot bodies are made up of vegetative strands of fungus.
- ☆ The sclerotic interior is white or tennis white in color.
- ☆ The size of grain kernel and ergot are similar.
- ☆ Its initial symptom shows honey dew occurrence during flowering stage.
- ☆ In this, fungus produces yellowish, sugary excretions which can be seen as droplets on flower parts.

Eyespot

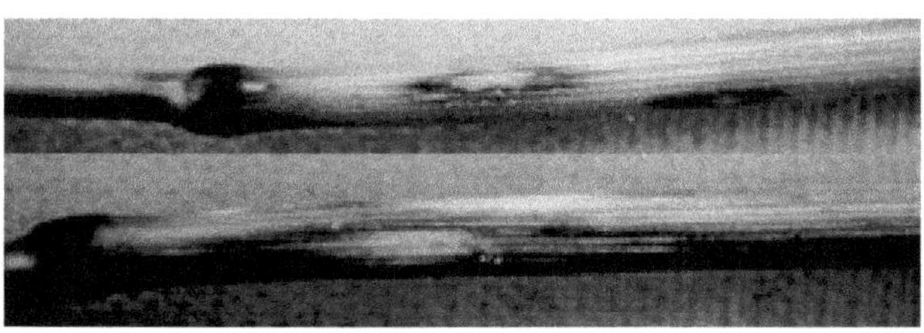

- ☆ It shows elliptical lesions which initially appear on leaf sheath and then spread to stem.
- ☆ In this, lesions are yellow-brown to tan in color and occur lengthwise till the stem.
- ☆ The lesions occur individually or in group which coalesces to form large areas of discoloration.
- ☆ The lesions girdle the stem and show a gray, thread like fungal growth on stem beneath the lesion.
- ☆ It causes mature stems to have charred appearance which infect tillers maturity at early stages and develop white heads and poorly filled seed.
- ☆ The tiller falls if stems are severely infected.
- ☆ It is mainly caused by splashing of water and high soil moisture content having dense crop canopy.

Fusarium head blight

- ☆ It forms one or more spikelets on newly emerged head bleached.
- ☆ It shows pink or orange fungal masses at the base of infected spikelet.
- ☆ The infected spikelets will not produce seed or shriveled and discolored seed.
- ☆ Its severe infection causes kernels to have chalky appearance and lost harvest.
- ☆ It is mainly in warm, moist conditions which help in spreading of disease.

Phythium root rot

- ☆ It makes the infected plants to become chlorotic and stunted.
- ☆ It helps in producing shriveled grain in plants.
- ☆ In mild infection, the tiller gets reduced with lowering of plant population and maturity.

Powdery mildew

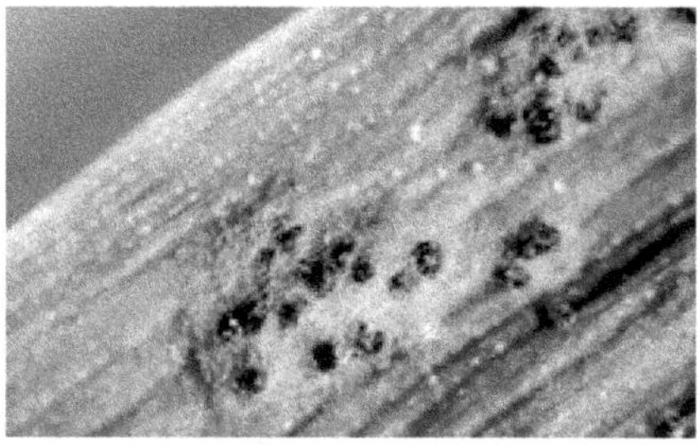

- ☆ It results in patches of cottony, white-gray growth on upper surface of leaves tha turn leaves to gray-brown.
- ☆ The chlorotic patches develop on leaves opposite fungal growth and make fungal fruit bodies as black dots on mildew.
- ☆ It lead to high nitrogen fertilization, high humidity and cool temperature.

Rusts/Stem rust, Leaf rust, Stripe rust

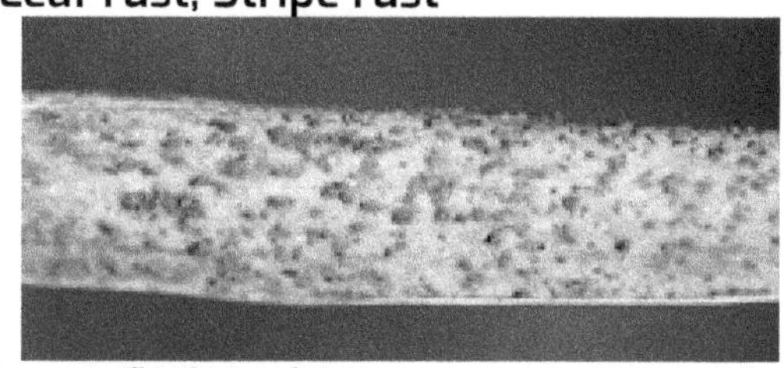

- ☆ It causes chlorotic flecks or brown necrotic spots on leaves or stems.
- ☆ In this, yellow streaks or patches on foliage appears which are brown which raises orange pustules as seen on lesions
- ☆ Its emergence favors cool and wet conditions.

Tan spot

- ☆ It is oval or diamond shaped necrotic lesions having brown centers with yellow halos on leaves.
- ☆ The infection requires wet period of 6-48 days.

6.4 INTEGRATED CONTROL MEASURES

- ☆ Preventative measures taken before planting or before the pest appears, can result in less rescue treatments.
- ☆ Each crop and situation will require management options tailored to that situation.

The control measure will include:

- ☆ cultural
- ☆ biological
- ☆ chemical
- ☆ genetic
- ☆ physical

Cultural Controls

- ☆ Rotate crops to reduce buildup of weeds, disease and insect pests.
- ☆ Crop rotation is useful for those pests that do not move far from their overwintering sites.
- ☆ Leave overwintering sites like cull piles, damaged and volunteer plants and alternate hosts, to minimize damage by insects and diseases.
- ☆ Use techniques that expose pests to natural enemies or environmental stress, or that makes crop less susceptible to insects or diseases.
- ☆ Adjust planting times to avoid periods of peak pest abundance.
- ☆ Promote vigorous crop growth with proper nutrition and weed removal to avoid stress that weaken crops and make them more susceptible to attack by insects, diseases or physiological disorders.
- ☆ Manage irrigation schedules to avoid long periods of high relative humidity as wet, highly humid conditions encourage disease pests to develop.
- ☆ Arrange fields for best air drainage and circulation to promote low humidity.
- ☆ Where crops are planted in rows use cultivation, where practical, in combination with banding of herbicides over row for weed control as it reduces herbicide costs with good weed control.

Physical Barriers

- ☆ It covers netting over small fruits and screening in greenhouses which helps insects that cause crop loss, and mulch can inhibit weed germination beneath desirable plants.
- ☆ These are important in termite, house fly and rodent control.

Biological Controls

- ☆ Conserving or releasing natural enemies can prevent rise of certain pests which includes mites that feed on mite pests in orchards, milky spore disease that kills harmful soil grubs, and Encarsia formosa, a wasp that parasitizes the greenhouse whitefly.
- ☆ Purchasing and releasing predators and parasites of pests, will be effective in reducing pest populations.
- ☆ Develop refuges for natural enemies of pest by creating areas of flowering plants and shrubs to supply nectar, hosts and shelter.

Pheromones

- ☆ It is a natural insect scents which can be used in pest management.
- ☆ By copying pheromone which female insect emits to attract mates can give confusion to males which prevent mating that curbs damage from different pest species.
- ☆ Pest-resistant cultivars are less susceptible than other varieties to certain insects and diseases.
- ☆ Planting disease resistant crops is simplest methods of reducing disease management actions during growing season.
- ☆ Use of resistant varieties will help farmers in growing wheat by keeping many pests at bay.

Chemical control

- ☆ Use pesticides when monitoring, economic thresholds or disease forecasts as per need.
- ☆ Select pesticides as per efficacy, previous use patterns, potential for and incidence of resistance, and possible impact on environment and natural enemies.
- ☆ Be certain to achieve uniform spray coverage with equipment by using recommended spray rates with correct calibrated equipment that targets the pest or protection of crop surfaces.

6.5 USE OF RESISTANT VARIETIES

- ☆ A variety is resistant if the pest cannot eat it, or cannot live or multiply on it.
- ☆ If a plant is resistant against a pest, it will not suffer any damage from that pest.

- ☆ Resistant varieties will not interfere with other pest-management techniques like crop rotation or, if needed, an insecticide application.
- ☆ These are only resistant to one or a limited number of pests.
- ☆ When resistant crops are being developed, some positive traits may be lost while other traits are strengthened.
- ☆ Some resistant varieties spend more of their resources protecting themselves; subsequently, their yields may drop.
- ☆ If resistant varieties are widespread and insect densities high, insects may develop new strains that are not affected by the plant's defenses.

Chapter 7

ABIOTIC STRESS MANAGEMENT

7.1 VARIOUS ABIOTIC FACTORS LIKE MOISTURE STRESS, TEMPERATURE, SALT ETC.

- ☆ The ability of wheat crops to tolerate abiotic stresses like excessive or inadequate supply of water, high winds, extreme temperatures, frost, salt, is main aspect of yield elasticity while its improvement has long target for cultivators and breeders.
- ☆ Due to experience to high temperatures in rainfed areas of globe, their exist drought stress.
- ☆ Prolonged high temperature may leads to osmotic stress as rapid evaporation of water from soils resulting in higher salt concentrations.

7.1.1 Temperature stress

- ☆ Wheat as a cool season crop which has optimal growing temperature during reproductive stage of around 15°C and above this, it shows reduction in yield of around 3%–4%.
- ☆ With increase of average temperature at 0.18°C per ten years, the impact of heat stress on wheat recently attracted the attention of growers.
- ☆ Heat stress causes injury to cellular structure and disturbs various metabolic pathways, specially photosynthesis, membrane thermostability, and starch synthesis related pathways.
- ☆ Denaturation of proteins and increased levels of unsaturated fatty acids due to heat stress disrupt water, ion and organic solute movement across the membranes, which shows increased cell membrane permeability and inhibition of cellular function.
- ☆ Temperature can modify developmental and growth rates in plants.
- ☆ Heat stress affects agronomic traits at all growth stage, while before flowering stage and anthesis period, there appears sensitive to high temperature after flowering stages.
- ☆ It is seen that, a high significant correlation exist between leaf chlorophyll content and grain weight.

7.1.2 Water stress

- ☆ Water is an important factor for wheat crops whose deficiency greatly reduces the yield.
- ☆ Many physiological processes in plants are impaired by drought stress, including photosynthesis, enzyme activity, membrane stability, pollen viability and ultimately growth.
- ☆ Drought is the main cause of yield and quality failure in wheat crops throughout the world's wheat grown area and developed countries.

- ☆ The drought resistance successfully achieved by avoiding or tolerating stress.
- ☆ In avoidance of drought, plants decrease the leaf area and increases canopy resistance, closing of stomata and formation of root hairs with mechanisms of tolerance at cellular and metabolic level.
- ☆ Resistance and tolerance mechanisms mainly focus on turgor protoplasmic resistance, maintenance and dormancy.
- ☆ Under water stress conditions, crop modify their biochemical pathways by changing of genes.
- ☆ Drought stress is along certain environmental stress factors, such as temperature, high solar radiation and wind.
- ☆ Water and nutrition are two main components of environmental variations and both limits successful crop production.
- ☆ Mineral nutrients are required for plant growth and development by their fundamental roles in plant metabolism, while drought is special among most important ecological factors that impact crop growth and productivity.
- ☆ Zinc, B and Mn are used in many physiological process in plant cell and many of these are linked with tolerance to drought stress.
- ☆ The nutrients play main role in maintaining photosynthetic activity, pollen viability, preservation of membrane integrity and continuance of enzyme activity.
- ☆ Many environments like arid and rainfed conditions shows early water stress in active growing season which causes less germination of wheat and crop establishment methods.
- ☆ Seed reserves are decreased showing less germination of plants and higher mechanical impedance of soil that reduces crop establishment.
- ☆ The negative effects of early water stress are less likely if bigger seeds of wheat are sown.
- ☆ The higher root biomass in newly grown seedlings from bigger sized seeds helps to maintain better water balance under early drought conditions and due to moisture in lower layers in soil profile.
- ☆ The active tillering stage sensitive to drought, which is half of biomass as compared to conditions if not in stress.
- ☆ The leaf area index stage is mostly affected stage during whole development of plant.

- ☆ Water stress before flowering or reproductive stages, leads to decreasing number of spikelet primordial.
- ☆ Under water stress, length increases but width has no effect, but the total leaf weight also remains same.
- ☆ Leaf extension is limited under drought for better absorption of water by roots and finally water conditions of plant tissues.
- ☆ For improving output in drought area, many steps are indispensable like incidence of water stress in environment and matching with phenology of crop matching with critical stages of soil moisture in particular climatic zones, so developing new means for better utilization of irrigation and increasing status of soil water with certain management practices.
- ☆ There are many field management practices to control water stress conditions like irrigation methods and identification of water stress resistance sources such as developing methods through screening.
- ☆ The repetition of experiments is important, but agronomic strategies as sprinkler irrigation system, water harvesting shelters and evaluating of drought susceptibility index are most important.
- ☆ The residue of wheat gets reduced with evaporation rate during growing season which slows the movement of water all through soil profile giving more time for water to infiltrate in profile.
- ☆ Crop rotation saves total water required by irrigation and during winter season, growing wheat lowers the irrigation requirements.
- ☆ The crop which is irrigated with 6 inches, leads to sowing of maize followed by wheat produced 8% higher the maize yield and following of maize in rotation.
- ☆ Crop rotation with irrigation facilities takes higher time than solo crop.
- ☆ Breeding the crops for water resistance, productions of higher biomass and water use efficiency are good ways of Agronomy.

7.1.3 Soil

- ☆ Abiotic soil environment influence the abundance and distribution of below ground insects.
- ☆ Soil drying prompts many root feeders to move down to deeper layers of rhizosphere.
- ☆ Abiotic factors can indirectly influence root feeders.

- The change in root physiology and metabolism, following abiotic stress may lead to changes in nutritional quality and defence capacity of roots.
- Abiotic factors influence the biotic environment that plants and herbivores encounter which includes natural enemies that may reduce herbivore pressure.
- Abiotic conditions influence plant environment interactions in quantitative manner where soil moisture ranges from flooding to complete dryness.
- The extreme conditions reduce both herbivore and plant fitness, unless organisms are specifically adapted to thrive under circumstances.
- The abiotic conditions fluctuate around levels that are in range of physiological compensation for both plants and insects.
- As abiotic changes are subtle, they strongly influence the outcome of plant herbivore interactions and their impact helps in explaining some of variability as observed in root herbivores.
- Certain conditions changes the relative fitness value of roots and consequently, defense patterns that plants should employ to defend their below ground organs.
- Water and nutrient limitation increases the importance of individual roots where under low stress conditions, plants can tolerate loss of significant part of root system without suffering any negative consequences.

7.2 PREVENTIVE CONTROL MEASURES FOR ABIOTIC STRESSES

Abiotic stresses include potentially adverse effects of:

- Salinity
- Drought
- Flooding
- Metal toxicity
- Nutrient deficiency
- High temperature
- Low temperature

In addition, abiotic stresses can include:

- Shade
- UV exposure Photoinhibition

- ☆ Air pollution
- ☆ Wind
- ☆ Hail
- ☆ Gaseous deficiency

Plants experience abiotic stress due to shortage of essential resource or from the presence of high concentrations of toxic or antagonistic substance. In cases like supply of water, too little (drought) or too much (flooding) can both impose stress on plants, so, abiotic and biotic stresses are often inextricably linked.

As the main production losses in wheat are caused by abiotic stresses factors like drought, salinity, and high temperature, so understanding such effects of stresses becomes indispensable for wheat improvement that depends mainly on genetic variations present in wheat genome by conventional breeding.

Plants resist low water potential and related stresses by modifying water uptake and loss to avoid low water potential, accumulating solutes and modifying the properties of cell walls to avoid the dehydration induced by low water potential and using protective proteins and mechanisms to tolerate reduced water content by preventing or repairing cell damage. Salt stress also alters plant ion homeostasis, and under many conditions this may be the predominant factor affecting plant performance.

Direct root

- ☆ Root traits plays a significant role in adaptive response of wheat crop to drought and low nutrients and their selection has often been advocated to mitigate yield losses in crops that are exposed to water and nutrient deficits.
- ☆ Breeders have largely neglected selecting for roots, not only for the demanding phenotyping, but also for difficulty in identifying yield-effective ideotype and to effectively select the desirable root architectural features.
- ☆ The root depth and root length density are mainly determined by basic (potential) root architecture and its development in time as plant develops and soil dries.

Temperature controlled

- In a method for phenotyping wheat genotypes for high temperature tolerance by integrating novel design of Temperature Controlled Phenotyping Facility with novel inexpensive tool to ensure uniform crop stand.
- In this, novel temperature controlled phenotyping facility allows screening of many wheat genotypes in larger plot size at desired temperature at any stage of crop growth, which allow plants to grow in natural environment at rest of the period.
- The motorized control units of it, allows roofs and walls to slide down to open the structure during initial growth stages which allow crop to receive prevailing environment conditions.
- During temperature stress treatment, structure is closed with sliding roof and windows so as to seal the whole unit.
- The temperature regulation in this, is precise and is linked to ambient temperature so that required difference between temperature in and out of structure is maintained with diurnal cycle during temperature stress treatment.
- For increasing temperature, boiler-based heating system is used where warm water runs through many pipelines that hangs from roof with many inlets and outlets to avoid formation of temperature gradient from one end to another end in structure.
- In this, cooling was done by split air conditioners, integrated and automatically governed by control panel.
- Mist system provides fine-water droplets to maintain required humidity levels, and the drip system for providing irrigation.

7.3 SELECTION AND USE OF RECOMMENDED RESISTANT / TOLERANT VARIETIES

Depending on the region where the farm is located, it will be subjected to different abiotic stresses. Acid soils and varieties having good low soil pH tolerance are great concern. While selecting wheat resistant crop, an important point is drought dominant. Varieties differ in their tolerance to abiotic stresses, and selecting a variety with better tolerance to the major limiting factor in operation will allow the variety's potential to be more easily achieved.

Selection and breeding of genotypes with improved drought/heat tolerance become key issues in the course of global change with predicted increased frequency of droughts or heat waves.

Drought sensitivity in crop varieties

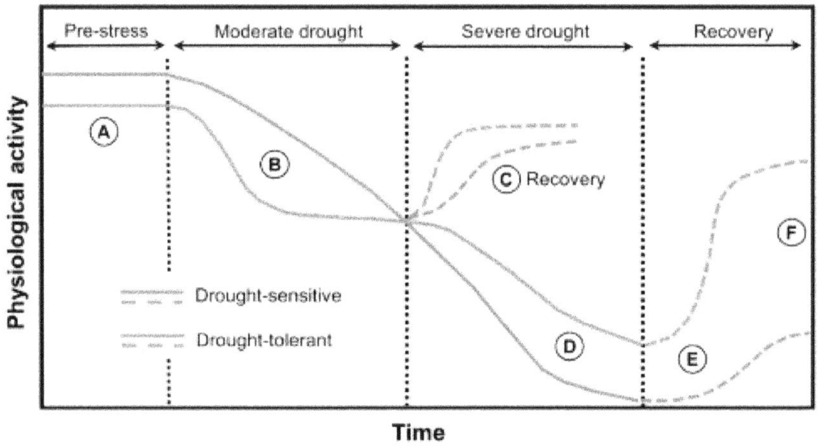

Fig genotypes differing in drought tolerance

- ☆ In case of non-stress phases performance is often lower for drought-tolerant than for well yielding standard varieties (1A).
- ☆ The performance during moderate drought decline more rapidly and remain stable on lower level in drought-tolerant genotypes, while standard varieties decline more steadily (1B).
- ☆ The down-regulation of physiological activities in stress-tolerant varieties improve the chances to survive by saving resources and serving as basis for rapid recovery after stress phase.
- ☆ In less drought tolerant varieties, there will be more productive at stress period, which then be more severely damaged and recovering less well than a more tolerant variety.

- ☆ An important point is velocity and degree of recovery after stress period (1C).
- ☆ The low activity level reached after severe drought (1D) where time course for physiological activities during subsequent recovery phase (1E) with completeness of recovery (1F) are important for genotype.

Chapter 8

QUALITY SEED PRODUCTION

8.1 CLASSIFICATION OF SEEDS

8.1.1 Introduction

Seeds can be recognized as:

- ☆ Nuclear seed
- ☆ Breeder seed
- ☆ Foundation seed
- ☆ Registered seed
- ☆ Certified seed

Breeder seed

- ☆ It is the seed or vegetatively propagated material which is directly controlled by originating or sponsoring breeder or institution serving as basic seed for recurring increase of foundation seed.

Foundation seed

- ☆ It is the progeny of breeder seed.
- ☆ The seed stock handled to maintain specific identity and genetic purity, which may be designated or distributed and produced under careful supervision of an agricultural experiment station.
- ☆ The seed is source of all other certified seed classes either directly or through registered seed.

Registered seed

- ☆ The progeny of the foundation seed maintain the genetic identity, purity, approval and certification by certifying agency.
- ☆ It serves as a quality which is suitable to produce certified seed.

Certified seed

- ☆ It is the progeny of the foundation seed.
- ☆ The production is handled to maintain genetical identity and physical purity as per standards as specified for crop certification.
- ☆ It should have minimum genetical purity of 99%.
- ☆ Certified seed have progeny of certified seed , provided the reproduction does not exceed two generations beyond foundation seed and provided if certification agency shows genetic and physical purity, if not be significantly altered.
- ☆ In highly self pollinated crops certification of further generation is permitted.
- ☆ Certified seed produced from certified seed is eligible for further seed increase under certification, except in case of highly self-pollinated crops, where certification of one further generation is allowed.
- ☆ Certification tags issued once for certified seed not eligible for further seed increase under certification.
- ☆ For paddy and wheat , certified seed produced from certified seed is eligible for certification by NSC up to two generations from foundation seed - Certified seed (I) - Certified seed (II)
- ☆ For barley, garden pea ,ground nut, soyabean, certified seed produced from certified seed is eligible for certification up to 3 generations from foundation seed - Certified seed (I) - Certified seed (II) - Certified seed (III)

8.1.2 Wheat Seed

The determination of wheat classes is done based on research and skillful experts who identify varieties of wheat seeds based on visual characteristics.

It is noted that the classification of wheat seeds depends on knowledge of seed:

- ☆ size
- ☆ shape
- ☆ color
- ☆ texture

It is required that above parameters should be used by machine vision for reorganizing various objects from each other. By assessing the discriminating power of such features for particular identification of seeds, quality seed from many seed species can be developed effectively based on computer vision system.

Apart from colour and morphology of wheat seed species, the textural features are more effective for classification techniques. Identification of Indian wheat seed variety and their classification efficiency of machine vision shows effective textural features of bulk samples. For getting textural features from grayscale images, the images are converted from gray into six matrices called as:

Gray matrix

LBP (local binary pattern)

LSP (local similarity pattern)

LSN (local similarity number)

GLCM (gray level co-occurrence matrix)

GLRM (gray level run length matrix)

While studying wheat seed classification, certain studies shows the use of textural features as its color and morphological features are principally used for such application.

8.2 QUALITY PARAMETERS OF WHEAT SEED

8.2.1 Seed Quality

- ☆ Seed quality is the possession of seed with required genetic and physical purity which carries physiological soundness and health status. The major seed quality characters are:

Physical Quality

- ☆ It is the cleanliness of seed from other seeds, debris, inert matter, diseased seed and insect damaged seed.
- ☆ The seed with physical quality should be of uniform size, weight, and colour.
- ☆ The seed should be free from stones, debris, and dust, leafs, twigs, stems, flowers, fruit well without other crop seeds and inert material.
- ☆ Seed should be devoid of shriveled, diseased mottled, moulded, discoloured, damaged and empty seeds.
- ☆ The seed should be easily identifiable as species of specific category of specific species.

- ☆ Lack of quality character in seed will indirectly influence the field establishment and planting value of seed.
- ☆ The quality character of seed could be obtained with seed lots by proper cleaning and grading of seed after collection and before sowing / storage.

Genetic purity

- ☆ It is true to type nature of seed.
- ☆ The seedling / plant / tree from the seed should resemble its mother in all aspects.
- ☆ The quality character of seed is important for achieving desired goal of raising the crop either yield or for resistance or for desired quality factors.

Physiological Quality

- ☆ It is the actual expression of seed in further generation / multiplication.
- ☆ Physiological quality characters of seed have seed germination and seed vigour.
- ☆ The liveliness of seed is known as viability.
- ☆ The extent of liveliness for production of good seedling or ability of seed for seedling with normal root and shoot under favorable condition is known as germinability.
- ☆ Seed vigour is energy or stamina of seed in producing elite seedling which is the sum of all seed attributes that enables its regeneration under any given conditions.
- ☆ Seed vigour shows level of performance of seed or seed lot at germination and seedling emergence.
- ☆ Seed that do well at sowing are termed as quality seed and based on degree of performance in production of elite seedling, it is categorised as high, medium and low vigour seed.
- ☆ The difference in seed vigour is differential manifestation of deteriorative process which oocurs in seed before ultimate loss of ability to germinate.
- ☆ Difference in seed vigour is expressed in rate of emergence, uniformity of emergence and loss of seed germination.
- ☆ The viable seeds needs no germination but all germinable seed will be viable.
- ☆ The vigourous seeds will be germinable but all germinable seed need not be vigourous. Physiological quality of seed is achieved by proper selection of seeds used for sowing and by caring for quality characters during extraction, drying and storage.

- ☆ Seed with good vigour is preferable for raising a good plantation as the fruits, the economic come out are to be realized after several years.
- ☆ The selection of seed based on seed vigour is important for raising perfect finalize plantation.

Seed Health

- ☆ The health status of seed is the absence of insect infestation and fungal infection, in or on the seed.
- ☆ Seed should not be infected with fungi or infested with insect pests as these will reduce the physiological quality of seed and also the physical quality of the seed in long term storage.
- ☆ The health status of seed includes deterioration status of seed expressed by low vigour status of seed.
- ☆ The health status of seed influences the seed quality characters directly and warrants their soundness in seed for production of elite seedlings at nursery / field.

8.2.2 General quality standards

Physiological stage

- ☆ For quick emergence and a good growth vigour, seed should have been stored for at least 2.5-3 months and no longer than 5-9 months as per storage temperature.

Virus diseases

- ☆ Seed to be used for multiplication should not contain more than 2% virus, so during field inspection, tolerance level of 1% virus is used.
- ☆ Seed supplied to higher altitudes to farmers organizations which produces own seed or seed for the market, the tolerance for virus diseases should not be more than 2% virus.
- ☆ Seeds to be used for production of table potatoes, more tolerance levels should be applied.
- ☆ A virus percentage of 10% gives yield reduction of not more than 2-3%.

Earth and extraneous matter

- ☆ Not more than 2% by weight.

Dry and wet rot

- ☆ Dry and wet rot which includes Phytophthora infestans has around 1% weight, except if caused by Synchytrium endobioticum, Corynebacterium sepedonicum, Pseudomonas solanacearum.

Tubermoth

- ☆ Not more than 2% of the tubers attacked.

Common scab

- ☆ Over 25% of tuber surface affects 5%.

Rhizoctonia

- ☆ Over 10% of tuber surface affects 5%

8.2.3 Parameters

S.No	Parameters	Range-I	Range-II	Range-III
1	Moisture of Wheat (% by wt)	11.5-12.5	12.6-13.5	13.6-14.0
2	Foreign matter (% by wt)	Upto 1.5	1.6-3.5	3.6-4.0
3	Other food grains (% by wt)	Upto 1.6	1.7-6.0	6.1-8.0
4	Damaged grains (% by wt)	Upto 1.0	1.1-4.0	4.1-5.0
5	Other Wheat's (% by wt)	Upto 5.0	5.1-20.0	Over 20
6	Slightly damaged grains (% by wt)	Upto 2.0	2.1-6.0	6.1-10.0
7	Weevil led grains (% by count)	Upto 1.0	1.1-6.0	6.1-10.0
8	Immature Shriveled and broken grains (% wt)	Upto 2.0	2.1-10.0	Over 10.0
9	Uniformity in grain size	More uniform	Slightly less uniform	Less uniform
10	Luster	Normal	Medium	Poor

8.3 PROCEDURE FOR QUALITY SEED PRODUCTION

8.3.1 Seed Production Protocol

The seed production process/protocol starts with proper identification of seeds bearing:

- ☆ clear scientific name
- ☆ origin of seed distribution
- ☆ description about its production

The process/protocol for production of planting materials includes seeds from clean planting material source which covers:

- ☆ Field facilities and equipment
- ☆ Source of material

- ☆ Field requirement
- ☆ Field inspection
- ☆ Agronomic practices like isolation, rotation and negative selection
- ☆ Harvesting and handling
- ☆ Post-harvest treatments
- ☆ Storage and transport
- ☆ Quality standards for supplied product

8.3.2 Characteristics/quality of good seeds

- ☆ Seed quality is perhaps the most important aspect of seeds production, acquisition and use. A bad seed or seed lot is not only a waste of money, it is a waste of time and it brings mistrust among farmers, seed producers and seed sellers. Bad quality seeds affect the entire crop production value chain. Even more important is that bad quality seed with low germination power or count can discourage farmers from purchasing them especially where efforts are made for introducing improved seed varieties. There are four basic attributes of seed quality to consider:

1. Physical qualities of the seed in a specific seed lot
2. Physiological qualities observes germination and vigour of seeds
3. Genetic quality which relates to specific genetic characteristics of seed variety
4. Seed Health which refers to the presence of diseases and pests within a seed lot.

8.3.3 Factors responsible for quality seed production

The quality of seeds is considered as an important factor for increasing yield. The use of quality seeds helps greatly in higher production per unit area to attain food security of the country.

Seed quality is determined by number of genetic and physiological characteristics. The genetic component involves differences between two or more genetic lines, while differences between seed lots of a single genetic line comprise the physiological component.

The genetic factors that can influence quality include:

- ☆ Genetic make-up
- ☆ Seed size
- ☆ Bulk density

The physical or environmental characteristics include:

- ☆ Injury during planting and establishment
- ☆ Growing conditions during seed development
- ☆ Nutrition of the mother plant
- ☆ Physical damage during production or storage by machine or pest
- ☆ Moisture and temperature during storage
- ☆ Age or maturity of seed

High quality seeds are the result of good production practices, which include:

- ☆ Proper maintenance of genetic purity
- ☆ Good growing conditions
- ☆ Proper timing and methods of harvesting
- ☆ Appropriate processing during threshing, cleaning and drying
- ☆ Appropriate seed storage and seed distribution systems.

8.3.4 Process for quality wheat seed production

Techniques of quality seed production involves following aspects:

Technical Aspect

- ☆ It involves different techniques that are adopted for producing quality seeds.

Management Aspect

- ☆ It involves management of individual farmers by cooperatives and groups for securing required quality of seeds and stable sale of products.
- ☆ The basic knowledge for wheat seed production involves general scheduling for wheat cultivation and technical work as shown:
  - 1st Week of Oct to 2nd Week of Oct, work of land preparation is carried out where at initial stage, fertilizer applications like farm yard manure of around 50-60 DOKO (250-300 kg /Ropani) are used for first ploughing.
  - Last Week of Oct, work of seed treatment is carried out fungicides using captan or thiram with application of fertilizer to field and plough of land before sowing using urea of around 4 kg/ Ropani, DAP of around 5 kg/ Ropani and potash of around 2 kg/Ropani for second ploughing.
  - 1st to 2nd Week of Nov, sowing seed and pressing soil by 250–300 seed/ square meter (3.5 kg/Ropani).
  - Last Week of Nov. to 1st Week of Dec, work of crown root initiation with 1st top dressing by using urea as 2 kg/Ropani for first weeding and irrigation.

- 3rd Week of Dec. to last Week of Dec, work of three tiller formation stage for first pressing of plants.
- 1st Week of Jan. to 3rd Week of Jan, tiller development stage work which involves first field inspection, uproot and removal of diseased and off type plants for white grub, cut worm, leaf blight, yellow rust and aphids for doing second pressing.
- 2nd Week of Feb. to 4th Week of feb, work of plant elongation stage using second top dressing with urea 2kg/Ropani for second weeding.
- 2nd Week of Mar. to 4th Week of Mar, spike initiation stage for second irrigation.

Climate and soil condition

(A)Climate

☆ Wheat crop is major crop of hilly region; however, it is cultivated both in hills and Terai region.

☆ Wheat crop requires cold and dry weather conditions. The optimum temperature for germination is between 20 ºC and for vegetative growth ranges from 16-22 ºC.

☆ If the weather becomes cool and it dews, the wheat plants grow many shoots and give more grains.

☆ The wheat plants can be under attack of disease and pests if the weather is very hot and humid at the time of sowing seeds.

(B) Soil

☆ Wheat can be cultivated in various types of soil.

☆ Fertile loamy black soil and the soil with good water holding capacity is appropriate for wheat cultivation.

☆ It is better to cultivate wheat in a soil that can hold little more water even in the dry season.

☆ The soil of the field for wheat cultivation should not be more acidic and alkaline.

Selection of land

☆ Land selection shall be done in coordination with seed production Cooperative.

☆ For wheat seed production, clean, plain, and fertile and land with a good drainage facility should be selected, and consider the following things to select such plot:

- Seed production plot should be fertile land with easy facility of irrigation and drainage. Wheat crop generally does not need irrigation; however irrigation is necessary at tillering and flowering stage).
- Easily accessible for supervision or field inspection.
- The land selected for seed production should not be diseased in previous crop time or it should be less infested with diseases or pests.
- It is appropriate to cultivate in a block of land than in small plots of land for seed production. It is better not to cultivate wheat for seed production for two years if the loose smut disease appears in the land.
- It is necessary to maintain spacing between the plots of wheat grown for seeds and wheat grown for cereal food or wheat of different variety for seeds as per the certified standard. To maintain spacing, a small plot between the cultivated plots should be left empty or barren.
- Wheat crop is a self-pollinated crop; hence, there will be less problem of external pollination. However; the isolation distance should be maintained in order to avoid transfer of diseases.
- Three meters of isolation distance between the different varieties of wheat crop cultivated for seeds or for different purposes is necessary. Likewise, if any variety of wheat is infested by loose smut disease, a spacing of 150 meters is necessary between the plots.

Chapter 9

HARVESTING, THRESHING AND POST HARVEST MANAGEMENT

9.1 CROP STAGE FOR HARVESTING

- ☆ Harvesting is the first and major post-harvest operation for separation, processing and storage of grains.

9.1.1 Harvesting of Wheat

Harvesting is the operation of cutting, picking, plucking digging or combination of operations for removing crop from under/above ground and removing of useful part of fruits from plants.

- ☆ It can be carried out using:
- ☆ Slicing action with sharp tool.
- ☆ Tearing action with rough serrated edge.
- ☆ High velocity single element impacting with sharp/dull edge.
- ☆ Two elements scissors type action.

Wheat requires normally 110-130 days between sowing and harvesting, as per the climate, seed types and soil conditions. Knowledge about crop stages is also important to identify periods of high risk from the climate.

Harvesting of wheat crop started in March and expected to continue till mid June while in rain-fed areas, harvesting of wheat and other crops begins after April. At harvesting time, what stalks begin to bend over from the weight of kernels made the entire plant to became golden in color. After the wheat is harvested from the field, machinery separates the stalks and chaff away from the kernels which are often used in applications from mulch to animal bedding.

Germination of seeds is affected by temperature, water availability, oxygen, light and substrate, maturity of seed and physiological age of the seed. Many plants need different variable for successful seed germination which are linked with individual seed variety and ecological conditions of plant natural habitat. The minimum water content needed for wheat germination is 35% to 45% by weight. Air is composed of 20% oxygen, 0.03% Carbon dioxide and 80% Nitrogen and seeds of most plant species germinate well in and environment providing this mixture of gases. The optimum temperature for germination is 12 to 25°C .Specific seed often have a temperature range within which it will germinates. Winter wheat requires exposure to cold temperature to enable flowering which is termed as vernalization, serving as acquisition of a plant ability to flower in spring by exposure to prolonged cold of winter.

9.1.2 Identification of appropriate crop stage for harvesting on the basis of colour and moisture

Time of Harvesting

The time of harvesting plays an important role. The following are certain harvesting precautions which need to be taken care:

- ☆ Wheat crop can be harvested when the grains become hard.
- ☆ Harvesting before maturity led to low recovery of grain, higher proportion of immature seeds, broken and poor quality and disease prone during storage.
- ☆ Delay in harvesting led to shattering and spillage of grains which results in exposure to birds, rodents, insect and pest attack.
- ☆ Harvesting to be done in dry summer season by using proper method and improved equipments.
- ☆ Harvested wheat should be kept separately to avoid mixing of varieties.
- ☆ Direct sun drying and excessive drying should not be done.
- ☆ Threshing and winnowing should be done in the fields. The grains should be packed in sound clean gunny bags to minimize the losses during transportation.

Harvesting grain

- ☆ Crops can be harvested at physiological maturity or at harvest maturity or storage maturity depends on situation or need of produce.

Physiological maturity

- ☆ It shows developmental stage where there will be no increase in dry matter in economic part.

- ☆ In cereals, moisture content of grains is very high at milking stage which deceases due to accumulation of photosynthates.
- ☆ There will be decrease in moisture content from 40% to 20% which shows physiological maturity.
- ☆ In this, translocation of carbohydrates is stopped due to formation of abscission layer between rachis and grain.
- ☆ In this, plant reaches maximum dry weight, increasing production inputs doesn't produce any gains in yield.
- ☆ The grains are of hard dough consistency.
- ☆ If grain is harvested before physiological maturity, it results in low dry matter, poor quality and will shrivel on drying.
- ☆ Various crops have different indicators of maturity.

The color of the grain gradually changes from green to golden brown. Grain ripens in three stages:

- ☆ Milky: The milk stage is the first part of maturing, and this is when the kernel begins forming. For this, press the grain and see that if milky liquid ooze out.
- ☆ Dough: In this, the liquid hardens inside the grain and the grain shows a dent if pinched. In this, kernel is fully formed and plant takes nutrients from stalk and leaves and transfers them to kernel. The kernel is still very moist at end of the dough stage which carries 30% water.
- ☆ Mature: grain is hard and the heavy heads often bend forward.

Harvest Maturity

- ☆ Such crop maturity appears when the product of interest is at best quality with maximum yield.
- ☆ For maturity, highest yield of produce is the best indicator for the farmers.
- ☆ In certain grains, harvesting is done around 25% moisture, while in many cases, harvesting of grain crops is done at 15 to 18% moisture.

Fig Harvest maturity

- ☆ If crops are grown for forage, best time for harvest is when crop has reached maximum vegetative yield with high quality.
- ☆ It takes place around 7-10 days after physiological maturity.
- ☆ The process involved in this includes loss of moisture from plants and grains.
- ☆ The symptoms of harvest maturity include yellowing of leaves, drying of grains or pods.
- ☆ In this crop harvesting results if there is need to vacate the field for sowing of other crop.

Storage maturity

- ☆ If there is no chance for post harvest drying, crop gets harvested at stage where it can be directly stored.
- ☆ In case of grains, <14% moisture content is allowed.

Crop Moisture Harvesting Range

The field should be monitored closely for harvest readiness. Moisture level determines when wheat is ready to harvest. Wheat is harvested in the summer months, after being harvested in the spring or winter. The moisture content of wheat is critical factor for determining when the wheat is ready for harvesting.

Harvesting of crop should be done between 20% and 14% moisture content.

- ☆ Physiological maturity occurs at about 40% moisture.
- ☆ Wheat may dry 2.5% per day between 20% and 40% moisture.
- ☆ The crop can quickly reach 20% moisture after it matures.

Harvesting above 20% moisture:

- ☆ Harvest damage to the kernels is more likely.
- ☆ Grain can be harder to store.
- ☆ Test weights can be lower.

Harvesting below 14% moisture:

- ☆ Cutter bar losses can increase.
- ☆ Test weights may decline each time the crop gets wet from dew or rain.
- ☆ Lodging may increase.
- ☆ Weeds may grow tall enough to interfere with harvest.

Harvesting at 18% to 20% moisture

- ☆ Allows earlier double-crop planting.
- ☆ Helps achieve maximum wheat and double-crop soybean yields.
- ☆ Helps maintain grain quality.

Effect of moisture on grain quality

Grains MC (%)	Effect of moisture on wheat
20 or more	Grains are soft in feel, can be pressed between teeth and started germinating if store for more than three days and results in less value as food or feed.
16 - 20	Grains may be pressed easily between teeth and do not germinate, gets attacked by moulds and bacteria, changes color, taste and nutritive value fast and are injurious when consumed
13 - 16	Grains can be crushed between teeth with pressure, not safe for storage due to loose lustier led to pest damage thereby developing sour taste with bad smell at high temperature which led to loss of weight at 3-5% month due to high rate of respiration.
10 - 13	Grains can be easily crushed and are not attacked by mould or bacteria but becomes susceptible to store insect pest. The pesticide measure is essential for safe storage.
8 - 10	Grains are hard to be crushed; resistant to insect attacks and retains weight, taste, germination and nutritive values for longer period. They gain weight due to absorption of water especially during wet season and are very safe for storage.

9.2 HARVESTING METHODS

9.2.1 Introduction

Harvesting of wheat is difficult process as it requires good preparation and timing. If dry wheat is left out in fields for longer time, winds and storms can destroy the crop. The quality of the wheat may decrease if wheat gets rained on which dries again. Harvesting of wheat requires use of combine- a heavy machine.

Harvesting of wheat started as soon as crop field gets dried enough so that it can be handled safely. For checking, use the moisture meter which will be useful in showing quick determination of crop condition. If meter is not available, then weigh 0.25- to 0.5-kg of sample, dry it at 260°F and reweight it again.

9.2.2 Methods

There are different methods of harvesting of wheat which are followed in different countries, as it mainly depends on timely labour availability and costs associated with harvesting. Harvesting of wheat crop can be done by:

- ☆ Manual
- ☆ Animal drawn machine
- ☆ Mechanically operated machine.

In wheat, the plants are straight and smooth and ears containing grains are at the top whereas most of oilseed and pulse crops have branches, which create problems' in harvesting by manual or mechanical means.

Manual Harvesting

Till 19th century, grain were harvested by cutting with sickle, scythe, cradle or manual mowers.

- ☆ It is the main method of harvesting in developing and under-developed countries.
- ☆ In developed countries, certain crops are harvested manually for high quality and premium price.
- ☆ In this, the crops are handpicked without using any tools.
- ☆ The sickle, spade and digging rods are main traditional tools for harvesting and digging while other tool which is used for harvesting is knife where output is low and effort required is high.
- ☆ In inderminate crop, harvesting is done at intervals as economic product maturity results at different periods.

Sickle

- ☆ Use of Sickle
- ☆ It is one of the common hand tools used for harvesting of crops.
- ☆ It consists of curved steel edged blade fixed to wooden handle.

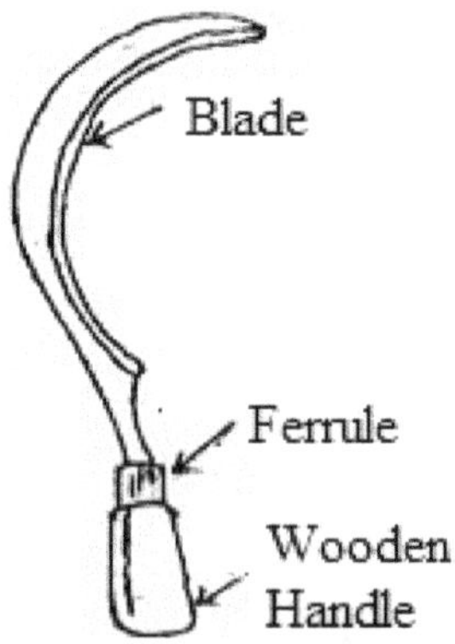

- ☆ The cutting edge can be plain or serrated, depending on its design.
- ☆ The curvature of sickle blade differs in plain edged and serrated sickles.
- ☆ The sickle blade is fabricated from medium carbon or high carbon steel.
- ☆ The blade has a knife section where its thickness reduces towards the cutting edge.
- ☆ The tang of the blade is inserted in wooden handle for joining the blade to the handle.
- ☆ The wooden handle is straight or bent at the end.
- ☆ In this, cutting is done by translatory and rotary movement to the blade around the point of cut.

Scythe

- ☆ It is a manually operated having a shaft of about 170 cm long known as snaith made of wood or metal.
- ☆ Snaiths are straight with offset handles having one or two short handles at right angles to it, usually one near the upper end and always another roughly in the middle.

- ☆ In this, handles are adjustable to suit the user.
- ☆ A curved, steel blade between 60 to 90 cm long is mounted at lower end at 90° or less to snaith.
- ☆ In this, the blade is projected from left side of snaith when in use, with the edge towards the mower.
- ☆ The mower holds the top handle in left hand and central one in right, with the arms straight with blade parallel and close to the ground and uncut wheat to the right.

Cradle

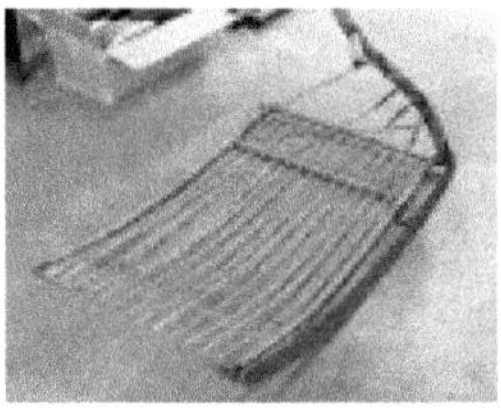

- ☆ It is a more developed tool for reaping and harvesting.
- ☆ It comprises of arrangement of fingers that are attached by light frame work which snatches of scythe.
- ☆ It gather straws as it is cut and deposit it in a swath.
- ☆ After cutting stroke, cradle is tilted to drop the straw in a pile.
- ☆ It can be useful in cutting more grains in a day.
- ☆ In this, due to more shattering loss, sickle is used to harvest the grains.

Reaper

- ☆ Reaper is a harvesting-machine used for cutting grain-crops.
- ☆ It delivers grain to one side in gavels which can be ready to bind into sheaves and/or elevate the gavels on platform where two operators bind them into sheaves by hand.

- ☆ Mechanical reapers and binders were developed to cut and windrow grain for field drying.

Self propelled vertical conveyor reaper

- ☆ It carries crop row divider, star wheel, cutter bar and lugged canvas conveyor belts.
- ☆ It cut the wheat crops and conveys it vertically to one end and windrows the crops on ground.
- ☆ It is a vertical type conveyor reaper which is walking type, self-propelled riding type and tractor mounted type.

Mechanical harvesting machines

Harvesting of grains by machines is an important part of mechanized agriculture. Power operated machines can be reaper, mower, rotary power scythes, forage harvester, binder and conventional combines.

The machines used for efficient harvesting of wheat crops include:

- ☆ Reaper binders
- ☆ Combine harvesters
- ☆ Tractor front and rear mounted reapers

1. Reaper binders

- ☆ It comprises of cutting unit as disc type or cutter bar type.
- ☆ After cutting, crop is conveyed vertically to binding mechanism which is released to ground in bundles.

- ☆ It is basically used for different kinds of low stem crops, rice and wheat.
- ☆ It can be used in different types of land forms like hills, slops, mountains, etc.
- ☆ In this, steer control handle bar is freely adjusted to up, down, left or right.
- ☆ It has a shaft drive system which works more stable and safe.
- ☆ It can be used on mountains or areas where combined harvesters cannot easily work.

2. Combine Harvester

- ☆ It is a machine which harvests grain crops such as wheat, oats, rye, barley, corn (maize), sorghum, soybeans, flax (linseed), sunflowers, peas, beans, lentils and canola.
- ☆ It it is a combination of harvesting—reaping, threshing and winnowing.
- ☆ It carries hydrostatic drives.
- ☆ In this, cutter bar runs the full length of header below the reel where its teeth gets open and close repeatedly to cut the crops from its base.
- ☆ Here, behind cutter bar, cut crops get fed toward the center by spinning augurs which travels up conveyor to processing mechanism in main part of combine.
- ☆ In this, unwanted chaff and stalks, passes along conveyors toward the back of the machine.
- ☆ Although the greatest application of combines is in harvesting the small grains, corn and soybeans, these are also used for wide variety of small-acreage or specialty crops.

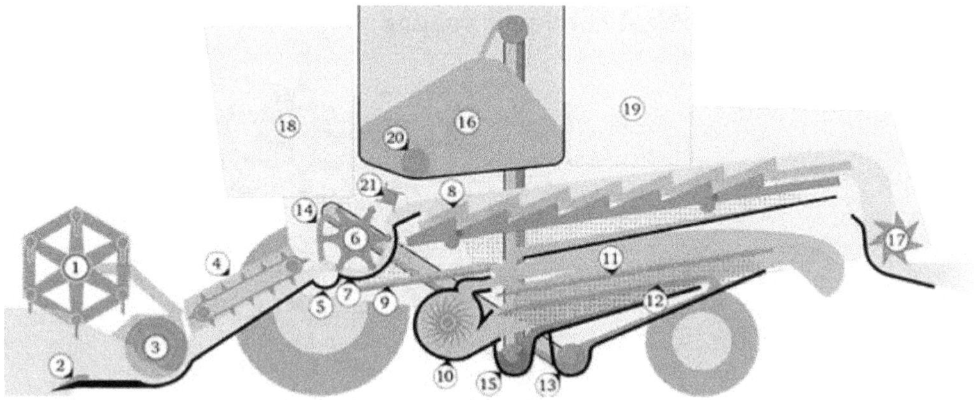

1) reel; 2) cutter bar; 3) header auger; 4) grain conveyor; 5) stone trap; 6) threshing drum; 7) concave; 8) straw walker; 9) grain pan; 10) fan; 11) top adjustable sieve; 12) bottom sieve; 13) tailings conveyor; 14) rethreshing of tailings; 15) grain auger; 16) grain tank; 17) straw chopper; 18) driver's cab; 19) engine; 20) unloading auger; 21) impeller

Its main function includes:

- ☆ Cutting the standing crops
- ☆ Feeding the cut crops to threshing unit
- ☆ Threshing the crops
- ☆ Cleaning the grains from straw
- ☆ Collecting the grains in a container

3. Small scale harvester

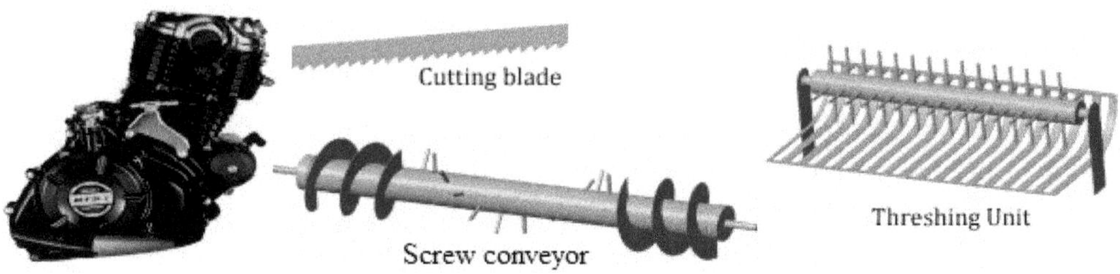

- ☆ It comprises of four stroke single cylinder petrol engine.
- ☆ It has a cutter blade assembly with sliding and stationery cutter blade where cutters are triangular in shape.
- ☆ In this, the screw conveyor collects the crops and transfers it for further operation.
- ☆ In this, the threshing unit threshers the dry crop which separates small grains from their chaff and straw.

4. Tractor mounted vertical conveyor reaper

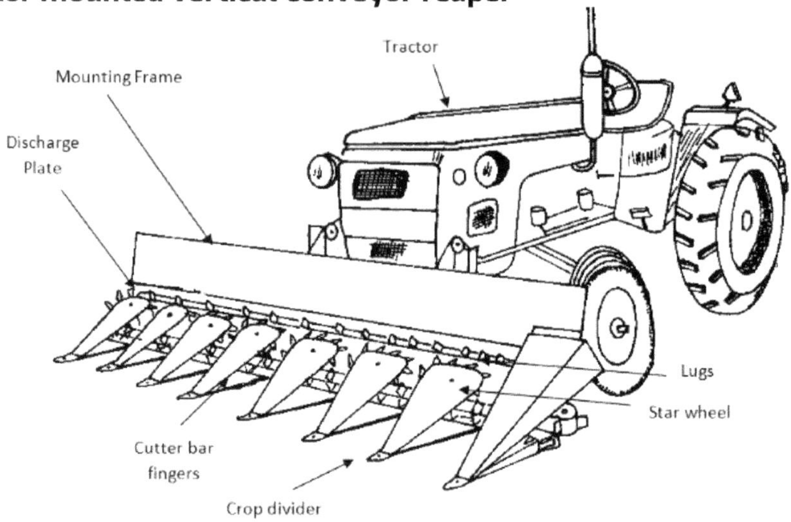

- ☆ It comprises of reciprocating cutter bar assembly, seven crop row dividers, two vertical conveyor belts fitted with lugs, pressures springs, pulleys and gearbox for power transmission system.
- ☆ In this, crop row dividers are fitted in front of cutter bar assembly.
- ☆ The star wheels are mounted over the crop row dividers.
- ☆ The machine is kept in front of tractor where the power is given from tractor PTO using intermediate shaft that runs below the chassis of tractor and coupling shaft.
- ☆ Here, the height of machine which is above he ground is controlled by tractor hydraulic using pulleys and steel ropes.
- ☆ Once the crop is cut by cutter bar, it is held in vertical position and delivered to one side of machine by lugged belt conveyors which falls on ground as windrow perpendicular to machine movement direction.

9.3 PROPER DRYING OF CROP BEFORE THRESHING

9.3.1 Introduction

After harvesting, wheat crop needs to be dried before threshing. If the harvesting of crop starts late, then the grain becomes too dry and rate of grain shattering is high. In such case, the moisture content of the grain will be high, which makes the drying difficult if harvesting is done early. The moisture content of wheat grain is an important factor from harvesting till milling. The moisture content of 25% is not uncommon in newly harvested grain in humid areas but it must be dried immediately to protect it against mould. At 14% moisture, grains can be safely stored for 2-3 months. For longer periods of storage from 4-12 months, the moisture content should be lowered to 13% or below.

9.3.2 Drying

Drying is a post-harvest system phase in which, the product is rapidly dried till it reaches safe-moisture level.

The idea of drying the grains is to lower the moisture content in it for safe storage and further processing.

Drying in wheat-growing countries is mostly done by spreading thin layer of grain in sun, on the threshing floor or on rooftops.

Mechanical drying of wheat grain is not practiced in many developing countries.

Wheat grains are mostly sun dried, since sun drying depends on weather conditions leading to dirty grain, spillage loss and bird attack.

9.3.3 Methods of drying

Drying of wheat crop is done in two stages. The first drying is by delaying the harvesting of the crop while second drying involves windrowed or bundled of harvested crop which is left in the field or transported to threshing site and stacked for further drying.

For drying wheat grain, normally two methods are used:

- natural drying
- artificial drying

Natural Drying

- In this, the grains are exposed to air (in sun or shade).
- For obtaining desired moisture content, grains are spread in thin layers on drying-floor, where it is exposed to air.
- In this, the duration depends on moisture content as required for safe storage.
- To get uniform drying, grains must be stirred frequently, especially if it is in direct sunlight.
- For effective drying, relative humidity of ambient air should not be more than 70%.
- Here, grains should not be exposed at night and the method should not be used in humid regions or during the rainy Seasons.

- ☆ The insufficient or excessively slow drying can led to severe losses in product during storage from self generated heat of green grain.
- ☆ Grains should be sun-dried in the middle of the day only and should not be dried in bags on a concrete floor.
- ☆ The process requires very low cost and has flexible capacity of grains o be dried up.
- ☆ It is a slow drying process which normally depends on the weather conditions.
- ☆ In this process, high risk of contamination with possible loss and risk due to birds and rodents.

Artificial drying

- ☆ In this, heated air (dryers) or unheated air (dehumidifiers) is blown through a grain mass.

Forced hot-air Drying

- ☆ This is the most wide spread practice in semi-humid and humid conditions where natural drying cannot be used.
- ☆ Artificially heated air is forced to flow through a mass of grain in bulk or in bags to absorb released moisture from grain mass
- ☆ It facilitates early harvesting and reduces shattering losses in the field
- ☆ It is more rapid drying in terms of capacity and is independent of weather conditions
- ☆ It requires heavy initial cost with high cost of drying
- ☆ In this, there is risk of seed damage due to high temperature
- ☆ It needs skilled personnel and regular monitoring with risk of breakdown or fuel shortage at critical times.

Dehumidified air-drying

- ☆ In this, unheated dehumidified air is circulated through the grain mass till moisture content of grain is reduced to required level.
- ☆ It is commonly applied when small quantities of grains are dried to very low moisture content for being used as seed.
- ☆ It is simple and easy to manipulate which involves low risk of seed damage from hot air
- ☆ It is not good for large-scale production as it involves high cost of investment

Types of dryers in artificial drying

For artificial drying of grains, two types of dryer are used :

Static or discontinuous dryers or batch dryers

- ☆ Such dryers are less expensive and can handle medium quantities of grain.
- ☆ Floor dyers : Suitable for all grains of all sizes.
- ☆ Bag dryers : Suitable for small grain lots to be used as seed where many varieties are handled (breeder and foundation seed). All types of grains in sacks.
- ☆ Box dryers: Suitable for large grain lot where slow drying is required.

Continuous dryers

- ☆ These are high-flow dryers that need a more complex infrastructure, complementary equipment and special planning.
- ☆ They are appropriate for big centers, silos or warehouses, where very large quantities of product are handled.
- ☆ Continuously flowing vertical dryers: Good for quick and normal drying and free flowing grains of single variety which can handle large volume of grain and are not good for small lots of different varieties.
- ☆ Continuously flowing belt dryers: Good for quick and normal drying seeds, small to very small lots and chaffy seeds.
- ☆ Rotary dryers: Good for quick and normal drying grains, chaffy grains with moderate moisture which are normally for large capacity drying and not for grains with high moisture.

9.4 PRECAUTIONS WHILE THRESHING AND SEPARATE THRESHING OF DIFFERENT VARIETIES TO AVOID MECHANICAL MIXTURE

9.4.1 Threshing

- ☆ Threshing is normally done after the grain moisture content is reduced to 15 to 17%.
- ☆ If harvesting of crop is done using combine harvester then there is no need for threshing the grains as the output is already in from of threshed grain.
- ☆ If harvesting is done by any other way, then threshing can be done by power operated thresher, threading head of wheat using bullock cart, beating grain with bamboo sticks, running tractor over the heads.
- ☆ Threshing is separation of grains from the plants which can be done in the field or on threshing floor by hand or with animals or machines.

- ☆ Threshing of wheat crop is manually fed into threshing cylinder and much of grains are removed from ear heads, cobs or pods when slow moving material comes in contact with high speed rotating cylinder rasp-bars.
- ☆ Threshing of crop takes place by agitation and rubbing action, as the material passes through the space between the cylinder and concave bars.
- ☆ Threshing efficiency depends on peripheral speed of cylinder, cylinder concave clearance, type of crop, moisture content of crop, weather condition and feed rate.
- ☆ Threshing is done by impact of rotating pegs that are mounted on the cylinder on the ear heads, which forces the grain out from the sheath that holds it.
- ☆ In wheat crop threshing, straw is bruised and broken up by the impact which converts it into straw.

9.4.2 Thresher

- ☆ Thresher was tractor operated type and used mainly for wheat. It was a very good machine, which threshed, cleared and bagged the grain, at the same time it made the quality straw (bhusa).
- ☆ Threshers are available in various sizes operated by 3-40 hp power sources.

Thresher comprises of:

- ☆ Drive pulley
- ☆ Fan/blower
- ☆ Feeding chute

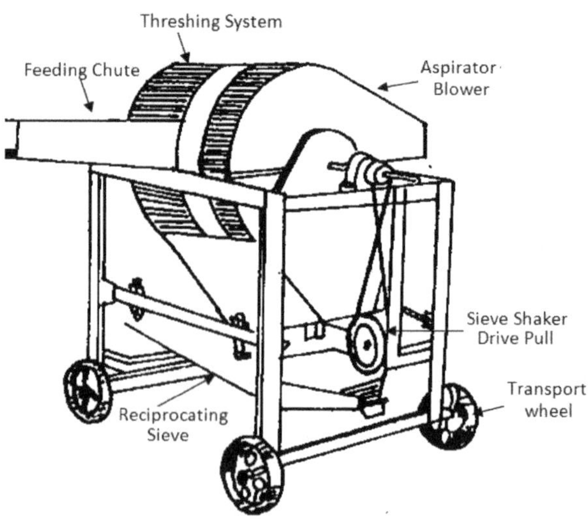

- ☆ Spikes
- ☆ Cylinder
- ☆ Concave
- ☆ Flywheel
- ☆ Frame
- ☆ Towing hook
- ☆ Upper sieve
- ☆ Lower sieve
- ☆ Transport wheel
- ☆ Suspension lever
- ☆ Can pulley
- ☆ Shutter plate

9.4.3 Precautions in Operating a Thresher

- ☆ Thresher machine should be installed at leveled surface.
- ☆ The direction of thresher machine should be set in direction of prevailing wind.
- ☆ Thresher should be operated at required speed as recommended by the manufacturer.
- ☆ It should carry suitable size pulley to operate the cylinder at required speed.
- ☆ In this, feeding should be continuous and uniform.
- ☆ Bearings and other working parts should be properly greased and oiled.

9.4.4 Threshing losses

- ☆ If the threshing of crop is not done properly, it could lead to following losses:
- ☆ Heaping of moist stalks in threshing floor lead to discolouration of mold, fungi attack, overheating, loss of viability, high free fatty acid etc.
- ☆ Damaged to kernel or pods will reduce market value.
- ☆ Delayed in threshing exposes produce for longer time on threshing floor which makes it vulnerable to damages from stray cattle, rodents, birds, weather hazards, grain loss in cracks.
- ☆ As grains are resilient in 11-13% moisture level which are capable of sustaining mechanical action with least damage, so under optimum conditions, threshing loss is kept less than 1-1.5%.

9.4.5 Avoiding Mechanical mixtures

- ☆ It is an important source of variety deterioration at the time of seed production.

- ☆ It results at the time of sowing when more than one variety is sown with same seed drill using volunteer plants of similar crop in seed field or by different varieties in adjacent fields.
- ☆ The seed produce of all the varieties are kept on same threshing floor, resulting in considerable varietal mixture.
- ☆ Avoiding of mechanical contamination can be done by rogue the seed fields and practice the utmost care at the seed production, harvesting, threshing and further handling

9.5 PROPER PACKAGING OF CROP PRODUCE IN CLEAN AND DISINFECTED BAGS

- ☆ Food grains like wheat and paddy should be stored in special bags.
- ☆ After threshing, the grains are separated from chaff or straw by lifting and tossing the threshed material so that lighter chaff and straw gets blown to one side while heavier seeds fall down vertically. Using hand-held winnowing baskets, the seeds are shaked which separates dirt and chaff.
- ☆ Before grains are kept in bags for storage or processing, they needs to be dried which can be done by spreading grains out in sun or with artificial dryer.
- ☆ Wheat grains after drying are stored in clean and disinfected bags for further processing. The grains should be inspected regularly for signs of spoilage and the moisture content tested.
- ☆ If the grain has moisture, it should be again dried.
- ☆ Once the grains are fully dried, they should be protected with insecticides and should be stored in rodent proof containers.

9.6 MAINTENANCE OF PROPER MOISTURE CONTENT BEFORE STORAGE

9.6.1 Introduction

- ☆ One of the most critical physiological factors in successful wheat grain storage is moisture content of crop.
- ☆ High moisture content leads to storage problems as it encourages fungal and insect problems, respiration and germination.
- ☆ Moisture content in growing crop is naturally high and only starts to decrease as the crop reaches maturity and grains are drying.
- ☆ In natural state, seeds have a period of dormancy and germinate when re-wetted by rain or by natural adequate moisture content.

- ☆ The moisture content of wheat crop should be checked by using formula; Weight of moisture/Weight of wet sample x 100.
- ☆ Grain are normally harvested at moisture content of 18–25% (wb), although it can be substantially higher or lower depending on maturity, season, weather pattern and drying facilities.

9.6.2 Wheat Equilibrium Moisture Content

- ☆ For drying wheat, high flow rate of air is forced though grain bed.
- ☆ The air is used as a medium to carry away moisture from the grain.
- ☆ The air temperature and relative humidity shows how much moisture the air can hold and with airflow, rate determines drying time, drying costs, final grain moisture content.
- ☆ Air at specific temperature and relative humidity passes through wheat which causes wheat to dry to specific moisture content as wheat moisture will eventually achieve a state of equilibrium with air.
- ☆ Temperature and relative humidity properties of drying air shows wheat final moisture level.

The recommended moisture content for safe storage of wheat grains are shown in table:

Storage duration	Required MC for safe storage	Potential problems
Weeks-few months storage	14% or less	Molds, discoloration, respiration loss, insect damage, moisture adsorption
8 - 12 months	13% or less	Insect damage
Storage of seeds	12% or less	Loss of germination
Storage for more than year	9% or less	Loss of germination

9.6.3 Keeping Grain Moisture Level Balanced

For keeping grain moisture level low/maintained, following methods should be applied:

- ☆ Before storing grains, it should be properly dried.
- ☆ Store houses should be well ventilated.
- ☆ Maintenance of adequate free space (1 1/2-2 ft.) between stacks.
- ☆ Stacks should be well separated from the wall and ceiling.

- Time to time aeration of store house reduces moisture content and prevents pest growth.
- Damp and wet bags should not be used for storing grains.
- Grains should be frequently inspected at least once in a month during dry season and once in fortnight during rainy season.
- The damp and moist grains should be disposed off at earliest or should be removed from the store house to be dried in sun.

9.7 STORAGE

9.7.1 Introduction

- Nearly two-third of total produce of wheat is retained at farm level for food, feed and seed.
- Majority of farmers uses traditional type of storage which is open to damage and loss from various sources.
- Wheat grain storability depends on:
    - moisture content
    - temperature
    - initial condition
    - insects and molds
- The main problems with existing facilities is short life of material used, not safe from weather effect, rodent and inadequate capacity.
- Majority of cultivators in all holding size classes, do not adopt improved storage facilities due to lack of funds, non-availability of desired type of container and in some cases unawareness about improved storage facilities.
- The farmers are aware of good quality storage structures and the precaution to be taken during storage as a of which, farmers adopted 'metal bin' for storing grains and oilseeds.
- Many farmers are following the method of open heaping and gunny bags for storage of produce.

9.7.2 Method

For storage of wheat grains, the store must satisfy the following parameters:

- grain must be kept dry
- grain should be kept at uniform temperature

- ☆ grain should be protected from insect attack
- ☆ rodents and birds should be excluded

The mode of storage in dwelling houses is tin drums, big earthenware, pucca kutcha kothas, thekas, underground mud walled or concretized cellars and R.C.C. bins.

Many indoor and outdoor structures are made of mud, wood, straw, jute bags, bricks and split bamboo.

The traditional storage structures varies from region to region which are not rodent, moisture and insect proof and require frequent maintenance.

Cereals are traditionally stored in mud kothi, jute bags in rooms or buried under wheat straw in underground structures.

Safe storage period reported from 2 months to one year.

There is a need to augment the production machinery to ensure appropriate and improved storage facilities like metallic bins, pusa bins etc are available to meet the needs of farmers in different holding size classes.

9.7.3 Disinfestations of store before storage of produce

- ☆ Before use of godowns/stores for keeping grains, it should be cleaned with dirt, rubbish, webbing and refuse materials.
- ☆ In godowns/stores, cracks, crevices, holes in wall, floor or ceiling should be filled with cement.
- ☆ Uneven floor, wall or ceiling in godowns/stores should be brought to level.
- ☆ The range of available insecticide products is broad enough to treat different surfaces as per their characteristics.
- ☆ It is essential to maintain the required concentration of fumigant in the sealed space and for sufficient time in order to have effective control of all stages of the pest.
- ☆ In case of uneven surfaces of godowns/stores, it should be treated by spraying with powder mixed with water.
- ☆ For smooth, non-porous surfaces of godowns/stores, spraying is done with stickier concentrate.
- ☆ Use of ambient treatment will help in destroying flying insects by aerosols in hermetically sealed premises which needs to be done mostly in evening when flying insects are active.
- ☆ Spraying do be done can be mechanical, pneumatic or thermal which provides better distribution of product over the grain.

Before storing of wheat grains, go-downs/store houses should be disinfected with methods of disinfestations as:

(a) Fumigation of empty store house with hydrogen cynide as it is widely used fumigant and is highly poisonous.

(b) DDT (dichloro/diphenyl/trichloroethane) and BHC (Benzene hexachloride) in form of smoke, dust or spray at concentration of 50-100 mg active ingredient per sq. meter. Excess use of DDT and BHC which are organo-chloride insecticides for stored grains produces health hazards.

(c) Malathion spray of around 100 mg/sq. m is better.

(d) Burning of sulphur.

(e) Good pyrethrum spray at about 6-8 oz/1000 c.ft is quite effective.

9.8 FUMIGATION OF STORE TO AVOID INFECTION FROM STORED GRAIN PESTS

Fumigation is a method of pest control which fills an area with gaseous pesticides to suffocate or poison the pests. It is used to control pests in buildings, soil, grain, and is used during processing of goods to be imported or exported to prevent transfer of exotic organisms.

Properly conducted fumigation will stop insect infestation and grain degradation from getting progressively worse. When fumigation is effectively conducted in late fall, pest populations can be drastically reduced.

Metallic drums or wooden boxes are used for fumigating small quantities of grain. Ethylene dichloride and carbon tetrachloride mixtures are recommended for fumigation of food grains in storage at farm level while hydrogen phosphide in form of aluminium phosphide or methyl bromide for protection in warehouses, godowns and silos.

Mixture of ethylene dichloride and carbon tetrachloride @ 1 litre for 20 quintals of grain or 35 litres per 100 m^3 of space with exposure period of 4 days is recommended. Methyl bromide is used at the rate of 3.5 kg per 100 m^3 of space with 10-12 hours exposure. The fumigant, hydrogen phosphide (aluminium phosphide), is available in tablet form and can be used at the rate of one tablet (3 g) per metric tonne or 25 tablets per 100 m^3 of space with an exposure period of 7 days.

For successful fumigation, following factors needs to be considered:

- ☆ Grains should always be level in store to allow fumigant to penetrate evenly.
- ☆ Any surface caking or crusting should be broken up and removed.
- ☆ Grain temperature should be 60°F or higher to ensure proper vaporization.
- ☆ Possible leak points like cracks or holes in store should be closed before fumigation as leakage will led to under treatment and poor control.

Chapter 10

BASIC FARM MANAGEMENT

10.1 INTERACTION WITH AGRICULTURE EXPERTS IN KVKS, KISAN MELAS, FIELD DAYS

10.1.1 Introduction

Wheat contributes to total food grain production of the country which is uptill 25%. Today, India is the second largest producer of wheat in the world. Wheat being major cereal crop of the country plays an important role.

For maintain the growth of wheat, considerable research in wheat cultivation are carried out. Wheat growers needs to regular update about the latest development which takes places related to wheat crop.

Fig Discussion with agriculture expert

Government has launched several farmer related awareness programmes. It is the duty of every wheat grower to attend those programmes and should meet different agriculture experts. During their meeting, agriculture experts will discuss on following with wheat growers which includes:

- ☆ role of weather and climate on crop productivity
- ☆ importance of weather forecasting in crop management
- ☆ farmer's problems
- ☆ advising wheat growers of suitable pest management practices in view of weather forecast
- ☆ information related to insects / diseases of wheat/other crops
- ☆ advices for protected cultivation and seed quality.
- ☆ benefits of weather based agro-advisory
- ☆ usage of insecticide, pesticide and polyhouse for raising healthy nursery
- ☆ emphasis on usage of FYM, green manures and crop rotation
- ☆ discuss about available forecast for particular crop
- ☆ information in new wheat research
- ☆ upcoming of new varieties
- ☆ weed challenges and wheat issues
- ☆ marketing

Fig Discussion with wheat growers

Wheat Growers can meet agriculture experts in:

- ☆ Krishi Vigyan Kendras
- ☆ Kisan Mela
- ☆ Field Day Trips

10.1.2 Krishi Vigyan Kendras

Krishi Vigyan Kendra was established in 1974 at Puducherry. In India there are 745 numbers of Krishi Vigyan Kendras in various districts. The Krishi Vigyan Kendra scheme is 100% financed by Govt. of India and Krishi Vigyan Kendra are sanctioned to Agricultural Universities, ICAR institutes, related Government Departments and Non Government Organizations in Agriculture.

The Kendras are interior part of National Agricultural Research System whose aim is to assess the location specific technology modules into agriculture and allied enterprises by technology assessment, refinement and demonstrations. They have been functioning as knowledge and resource centre of agricultural technology for public, private and voluntary sector in improving agricultural economy of district.

Fig Krishi Vigyan Kendra

Wheat growers/farmers visit these kendras and its instructional farm, demonstration plots/units and information centre for getting knowledge related to current agricultural practices & in solving their queries.

They conduct exhibitions on different themes for dissemination of new technologies, to increase their knowledge and interest about agriculture information/technique.

Activities of Krishi Vigyan Kendra

- ☆ Farm Advisory Services
- ☆ Training programme for different categories of people
- ☆ Training programme for extension functionaries
- ☆ Front Line Demonstration
- ☆ On Farm Testing

On-Farm Testing:

- ☆ KVK operates small farm to test new technologies like seed varieties or innovative farming methods.
- ☆ The ICAR institute allows new technologies to be tested at local level before transferring to farmers.
- ☆ It is an adaptive research which is conducted on farmer's fields by farmers with the support from scientists/subject-matter specialists.

- ☆ It is conducted with a farming systems perspective.
- ☆ The main idea is to find existing inputs of practices that might help in solving main problems of many farmers in defined study area.
- ☆ Farming system perspective implies seeing things from farmers viewpoints.

Front-line Demonstration:

- ☆ Due to KVK's farm and its proximity to nearby villages, it organizes programs to show efficacy of new technologies on farmer fields.
- ☆ KVKs organize front line demonstrations that shows the production potentialities of newly released and pre-released production technologies of cereals, pulses and oil seeds and farmers fields.
- ☆ In this, technologies are demonstrated for first time before fed into main extension system.
- ☆ The KVKs are given with responsibility of conducting good integrated farming system demonstration which serves as model for extension agencies.

Capacity Building:

- ☆ KVK hosts capacity building exercises and workshops to discuss modern farming techniques with groups of farmers.

Multi-sector Support:

- ☆ KVR gives support to many private and public initiatives by its local network and expertise which is common for government research institutes to leverage network of KVKs on doing surveys with different farmers/growers.

Advisory Services:

- ☆ With increasing use of ICT, KVKs implements technologies to give information to farmers on weather or market pricing by radio and mobile phones..

10.1.3 Kisan Mela

Krishi Vigyan Kendra organizes Kisan Mela to provide latest agricultural information, guidance, motivation to farmers and wheat growers to adopt new scientific and profitable practice. Farmers and wheat growers get benefited to know about latest technologies like improved farm machinery, new improved varieties, harvesting and management of crops etc.

Through Kisan Mela, farmers and wheat growers get interaction with scientists which will help in transferring of new agriculture technologies in different areas. Farmers and wheat growers gets acquainted with latest and recent developments in agriculture sector.

Purpose:

Kisan mela is a regular activity of agriculture department along with Baisakhi celebration in which local framers/crop growers and adjoining states farmers participates to understand the new inventions in area of agriculture.

Farmers/wheat grower participation in kisan mela and gathering information about latest recommendations made in agriculture will help them in enhancing their knowledge and understanding better crop economics return.

The purpose of organising kisan melas across the different parts of the country will help wheat growers/farmers in getting following information while discussing with experts which includes:

- educating farmers to stop burning paddy straws and focus on stubble management.
- information related to agro-commercial sector, agro based industries and export oriented market.
- making aware of technical know-how for cultivation of various crops.
- giving technical knowledge about the type of equipments used.
- identifying crop disorders in plants pertaining to insect and pathogen injuries, phytotoxicity of agricultural inputs or nutrient deficiency.

10.1.4 Field Day:

Field days are great way to share innovative ideas in sustainable agriculture with fellow farmers and wheat growers. It is an educational events which is hosted by producer or educator and held on-farm or on ranch.

It includes:

- demonstrations of specific management practices
- demonstrations of specific equipment

- highlighting research methods and results

In field day, the audience includes:

- fellow producers
- agriculture professionals
- students
- community members
- starters and established farmers/crop growers

The field day includes:

- Presentations
- Posters
- Materials

Farmer Field Days are production that focuses on-farm workshops to help farmers and ranchers in getting practical information which can be used to plan or enhance their operations. Farmer Field Days are geared toward those:

- who are just at start of their farm exploration–farm hands, interns, folks thinking about starting a farm in near future
- anyone in startup of crop production.

Field Days are arranged to show new technologies in front of large manageable group of interested farmers/crop growers. With this, farm experts, extension workers and farmers are involved who learn from each other.

10.1.5 Field visits

Field visits at different villages were conducted and feedback from the farmers was obtained. The farmers were enthusiastic in getting the agro-advisories, especially regarding the frost forewarning along with the precautions to be taken.

The idea behind field visit by experts is to:

- discuss with farmers where farmers asked several questions about different cycles of improved hybrid and dual purpose crop varieties which will help in making most out of the demonstrated technologies.
- learn more about the advantages and disadvantages of using chemical fertilizers, improved sorghum varieties, soil, water and crop management practices.
- increase farming productivity based on best performing varieties and fertility treatment.

10.2 MAINTAIN RECORD OF CROP PRODUCTION, INVESTMENT AND EXPENDITURE

10.2.1 Records

Keeping track of what is happening on the farm requires some records. Good farm management requires good useful set of farm records which does not ensure successful of farms, but success is unlikely without them.

Record-keeping is a task that has been started many times by eager farmers, only to be dropped later. Reasons that complete, accurate records are not kept revolve around lack of time and motivation. Keeping records is productive work.

Good record-keeping system should take into account the wheat growers needs, abilities and limitations. Wheat growers should need to understand the purpose for keeping the records and when the purpose is known, it will help them in deciding which records are useful and need to be kept.

With accurate facts and figures is useful while borrowing money, seeking government support and completing tax returns. From reliable farm records, equity proportion of total assets owned can be updated to assist with future farm investment programs.

Farm records can help wheat growers in evaluating wheat enterprise on mixed farm while farm advice will be more effective if adviser knows exactly what is happening on the farm. Records and statements show what has occurred or is occurring on the farm while budgets predict what might happen in the future.

Records provide wheat grower with data, information and knowledge. There are four uses for farm records such as:

- ☆ Use as service tool which covers income tax calculations, estate planning, business arrangement reconciliation and obtaining and managing credit.
- ☆ Providing data for financial analysis and diagnostic instruments like identifying strengths and weaknesses of business.
- ☆ Progress indicator where series of records are required to monitor work progress.
- ☆ Forward planning tool for keeping past records to project future cash flows for current and modified farm management practices.

10.2.2 Details of Record

Records of farm comprises of:

Farm diary

- ☆ It is a form of small notebook where wheat grower should record key facts and figures about farm business and its day-to-day activities.

- ☆ Farmers /wheat grower who use diary finds important facts and figures which could be easily lost or forgotten are stored permanently in this for future reference.

Crop records

It is useful to record what happens to each plot or crop each year that will includes:

- ☆ crop,
- ☆ fertilizer applications,
- ☆ agronomy (seeding, weeding, sprays),
- ☆ visual impression of crop,
- ☆ harvest interval of forage crops
- ☆ crop yield

Unit costs

It covers all main farm inputs like:

- ☆ fertilizers,
- ☆ fuel,
- ☆ irrigation water,
- ☆ concentrates and their ingredients,
- ☆ purchased forages,
- ☆ stock purchases

The details are required for routine book keeping and in monitoring seasonal changes and will help in planning future purchases.

Unit returns

- ☆ It includes returns from all farm outputs such as sale of crop, manure etc.

Crop stock accounting

- ☆ The main purpose of crop stock accounts is to monitor net losses and gains in income, and to distinguish the increases and decreases due to changes in market value.
- ☆ When quantifying total stock value, wheat grower has to take into account both in change size and change in unit of the crop.

Plant and improvement records

It covers records related to:

- ☆ date and cost of purchase or installation,
- ☆ annual depreciation,

- insurance and registration,
- fuel use,
- hours use,
- repairs and maintenance

Key financial records

- It covers detail related to interest and principal repayment schedules to plan repayments.

Other Farm Charges

It include records related to:

- rates and government charges,
- telephone and office upkeep charges,
- labourer's wages,
- capital investments,
- unusual weather events,
- important meetings,
- credit repayments,
- changes in valuation of capital

Personal expenses

- It includes expenses related to personal input wages which includes valuation of any farm produce used for personal consumption.

Different records

- It covers cost related to crops, fertilizer, seeds, spray, labour, water, machinery, harvesting, processing, transport and selling.
- In this, wages are broken down into different farm activities.

10.2.3 Importance of Records

- They are management tool which allows wheat growers to measure how efficiently they are using resources and to determine whether or not they are having income.
- The records are essential for planning and decision making
- The records are used in obtaining credit
- The records are used for income tax

10.2.4 Farm Accounting

It is the process of measuring and recording all farm resources and all farm transactions with financial consequences in a systematic way. The conduct of farm accounting takes large part of wheat grower/farmer's time and effort which gets compensated in account of following reasons:

- ☆ It permits farmer/wheat grower to find out the size of income which is derived from the farm.
- ☆ It helps in knowing the total value of farm business and to know which part is actually owned by him/her and which by others.
- ☆ It will help in detecting loss or theft of cash or stock.
- ☆ It provides necessary data for correct income tax assessment.
- ☆ It helps in claiming expenses.
- ☆ It helps in keeping complete information as farmers/wheat grower dislike paper work since they are busy with their farm work.

10.2.5 Farm Financial Records

Farm transactions can be recorded when they occur or at the time of cash transactions. Since farmers are involved in lot of cash transition, so they should use accrual method of accounting, which will help them in tracking of their money, inventory, sales etc. For example, when recording sales of wheat for specific time period, all the expenses needed to generate that wheat are relevant when calculating the cost of production.

Every physical item on farm can be placed into following categories:

1. Assets: It is required to generate income which can be either owned/purchased on credit that can be combination of physical and monetary values, where physical quantity is multiplied by unit price to have its value.
2. Liabilities: It relates to items on credit.
3. Equity: It serves as an asset with less liabilities.
4. Revenue: It is source of income that is generated from the business.
5. Expenses: It involves the costs what is required for a business to generate revenue.

Annual financial accounting comprises of:

1. Opening balance sheet: record all assets and liabilities at beginning of financial period.
2. Closing balance sheet: record all assets and liabilities at end of financial period.
3. Cash flow: record all cash inflows and outflows in business.

4. Income statement: calculate wheat enterprise profit/loss before or after income tax.

10.3 MAINTENANCE OF CROP CALENDARS

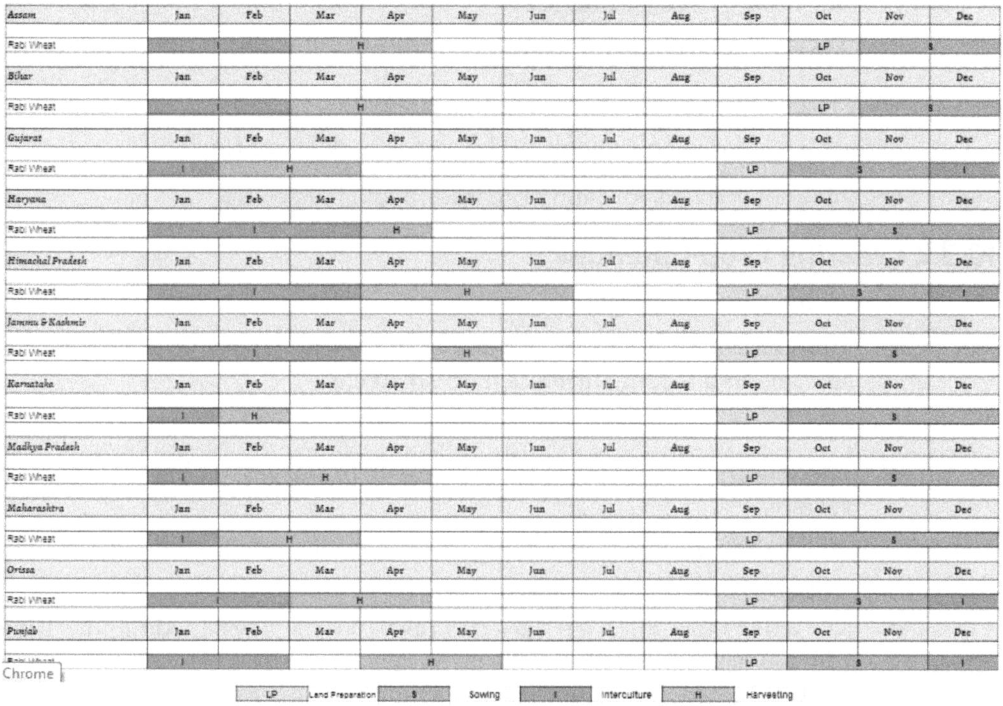

Fig Crop calendar

10.3.1 Crop Calendar

Crop Calendar is a form of tool which gives timely information about seeds to promote local crop production. It contains information on planting, sowing and harvesting periods of locally adapted crops in specific agro-ecological zones.

- ☆ It provides information on sowing rates of seed and planting material and main agricultural practices.
- ☆ It supports farmers and agriculture extentionists across the world in taking appropriate decisions on crops and their sowing period, respecting the agro-ecological dimension.
- ☆ It provides a solid base for emergency planning of rehabilitation of farming systems after disasters.
- ☆ It allows better planning of all farm activities and the cost of production.

- ☆ It serves as schedule of wheat growing season from fallow period and land preparation, to crop establishment and maintenance, to harvest and storage.

The crop calendar allows a farmer to:

- ☆ plan for input purchase and use
- ☆ develop cash flow budget for year
- ☆ determine need credit and period requirement
- ☆ determine labor requirements and plan for peak usage times
- ☆ organize contractors for land preparation and harvesting

10.3.2 Creating Crop Calendar

- ☆ Determine best date to plant:
  - This information can be gathered from local experience, agricultural advisors and leading farmers in the district.
- ☆ Determine time the variety takes from planting to harvest:
  - The length of time from establishment to harvest is known for each variety.
  - It may vary little as per growing conditions, especially water availability and solar radiation.
  - Normally short duration varieties take 100–120 days, medium duration 120–140 days, and long duration 160 days plus.
  - Most varieties take 60–65 days from panicle initiation to harvest.
- ☆ Mark on calendar, date of planting and when operation such as plowing, weeding, fertilizing, harvesting needs to be done.
- ☆ Determine the amount of labour, equipment and finance required at each step during the growing period.

Wheat crop calendars of weed, insect pest and diseases

10.4 AWARENESS ABOUT GOVERNMENT SCHEMES

10.4.1 Introduction

- ☆ Subsidy is an important fiscal instrument, apart from taxes and others by which, objective of growth and social justice are achieved.
- ☆ It is necessary as production accelerating catalyst for new inventions, that are socially desirable but whose adoption requires much capital and producers believe it as risky investment.

- ☆ It serve as right instrument which maximise risk taking.
- ☆ It serves as manipulating or balancing of growth rates of production and trade in many areas and regions and for equitable distribution of income for saving weaker sections of society.
- ☆ It acts as negative taxes, which serve as instruments that transfer resources in favour of those who receive them.
- ☆ It serve as direct or indirect, cash or kind, general or particular, budgetary or non budgetary, etc.

Wheat growers should be aware of various government schemes, which will help them in getting subsidies for buying implements, seeds, fertilizers etc. There are many governments, semi-government, state government schemes where farmers/ wheat growers can take subsidy for buying instruments and farm related products. Such schemes not only helped farmers to gradually adapt to new practices of farming but also aided in implementing green revolution by supporting farmer.

10.4.2 Agriculture Subsidies

Farmers/wheat growers should understand the following types of agriculture subsidies that are issued by various department in order to buy goods, that are categorized as:

Direct and Indirect subsidies:

- ☆ Direct Farm subsidies involve rendering cash to the recipient farmers.
- ☆ India provides direct subsidies in limited form like food subsidy, MSP-based procurement, etc.
- ☆ Direct farm subsidies are very common in most of developed countries like US and Europe.
- ☆ Indirect farm subsidies are not provided in form of cash but support farmers in indirect manner.
- ☆ Direct subsidy provides cash directly to farmers to buy fertilizers whereas indirect subsidies involves subsidizing of fertilizer companies for providing cheap urea to farmers.
- ☆ Direct subsidies are important as they provide direct purchasing capacity of farmers, thereby, having multiplier effect in enabling farmers to invest in agriculture and eventually increasing their standard of living.
- ☆ Direct subsidies help in checking the misuse of public funds as they help in the proper identification of beneficiaries.

- Direct subsidies gives choice to consumer to invest as per its need.
- In case of indirect farm subsidies, 2% is spend on GDP.
- Among all products, fertilizers accounts for two thirds of total indirect subsidies.
- Indirect subsidies are beneficial to big as well as small farmers which are more dependent on cash to meet their needs.
- Indirect subsidies do not provide incentive to farmers to invest in improving production processes.

Explicit Input Subsidies

- It shows the payments made to farmers so as to meet part of the cost of an input.
- It serve as the nature of explicit payments that are made to farmers.
- Such subsidies are made available to small and marginal farmers and those belonging to reserved categories.
- Input subsidies are induced to farmers to adopt yield increasing inputs so that they are able to realize the benefits of new technology.
- The subsidies will cover crops, inputs, regions and target groups from time to time.
- The subsidies have formed only a part of fraction of development expenditure of Central/State Governments.

Implicit Input Subsidies

- The implicit input subsidies are hidden in nature that appears on account of mechanics of pricing of inputs.
- If inputs whose prices are administratively determined are priced low as compared to their economic cost, acting as case of implicit subsidization.
- In this, farmers will not receive any direct payment but somebody in economy accounts for its difference.

Output Subsidies

- Subsidization of agricultural sector by output pricing shows the restrictive trade policy, product prices in domestic market that are maintained at levels higher than those that would have prevailed in absence of restrictions on trade.
- If trade policies have resulted in keeping domestic prices lower than

corresponding border reference price, then policies result in agricultural sector tax.

- The border reference price is free on board prices in terms of exportables and cost, insurance and freight price in case of importable.

10.4.3 Policy Instruments of Subsidy:

There are five main policy instruments that delivers direct and indirect farm subsidies:

1. Price and Income Support Policy

Government has many instruments to safeguard the interest of farmers/wheat growers under its price and income support policy which are:

- Minimum Support Price (MSP)
- Minimum Export Price (MEP)
- Market Intervention Price (MIP)
- Buffer Stocks Operations
- Public Distribution System

Minimum Support Price:

- In Minimum Support Price Scheme, CACP recommends MSP currently for 26 commodities.
- On calculating cost of production, commission considers the cost of paid-out inputs, imputed value of family labour and land rental.
- The price are normally announced before the commencement of sowing operations of crop.
- Such price are usually remunerative and higher than the production cost of crop.
- In this, farmers/ wheat growers are free to sell their product either at MSP or in market whatever is benefit for them.

Minimum Export Price (MEP):

- In this, government disallows exports of particular commodities below certain price.
- It shows checking of domestic price rise and augment supply in domestic market.
- It is used as temporary measure which help farmers/exporters to realize better and remunerative prices.

- ☆ Government generally applies MEP on price sensitive commodities like pulses, rice, edible oils, wheat etc.
- ☆ The price decisions are taken by union cabinet on request of ministries.

Market Intervention Price Scheme (MIP):

- ☆ It is a price support mechanism implemented by central government on request of State Governments for procurement of perishable and horticultural commodities in event of a fall in market prices.
- ☆ It is implemented when there is at least 10% increase in production or 10% decrease in the ruling rates over the previous normal year.

Buffer Stocks Operations:

- ☆ The government maintains buffer stocks of food grains (wheat) and other commodities (pulses) procured by agencies by MSP or other procurement schemes to check sudden price rise of commodities and any such other contingencies.

Public Distribution System:

- ☆ Distribution of subsidized food to poor is at the core of food security system.
- ☆ It is operated by TPDS and managed by FCI, who is responsible for procurement and buffer stocking of grains.
- ☆ Wheat/rice is distributed through TPDS to benefit the poor.

2. Research and Development Support

The government implements the policy and schemes to augment the public investment in R & D with focus on increasing production and productivity of farm commodities.

3. Input Subsidies

- ☆ The central and state governments provide input subsidies for farmers/ wheat growers by subsidized fertilizers, electricity, irrigated water and seeds.
- ☆ Also, commercial banks, co-operative societies and regional rural banks are required to provide credit to wheat growers below the market interest rates.

4. Import Measures

- ☆ Government imposes tariffs, quota, state trading and monitoring agencies as import measures to provide indirect subsidies to wheat growers.
- ☆ Government also imposes import duties on farm products to protect farmers/ wheat growers by preventing flooding of cheap agricultural commodities.

5. Export Measures

- ☆ It includes measures to increase farm exports.
- ☆ The main instruments under this include agri-export zones, incentives and price support on farm exports and establishment of various promotional councils.

10.4.4 Types of Agricultural Subsidy

Fertilizer Subsidy:

Disbursement of cheap chemical or non-chemical fertilisers among the farmers. It amounts to the difference between price paid to manufacturer of fertilizer (domestic or foreign) and price, received from farmers, rest of the burden is bear by the government. This subsidy ensures:

1. Cheap inputs to farmers,
2. Reasonable returns to manufacturer,
3. Stability in fertilizer prices, and
4. Availability of fertilisers to farmers in adequate quantity at the requirement.

In some cases this kind of subsidies are granted through lifting the tariff on the import of fertilisers, which otherwise would have been imposed.

Power Subsidy:

The electricity subsidies imply that the government charges low rates for the electricity supplied to the farmers. Power is primarily used by the farmers for irrigation objectives. It is the difference between the cost of generating and distributing electricity to farmers and price received from farmers. The State Electricity Boards (SEBs) either generate the power themselves or purchase it from other producers such as NTPC and NHPC. Power subsidy "acts as an incentive to farmers to invest in pumping sets, bore-wells, tube wells etc. Irrigations subsidy: under this umbrella government provides irrigation facilities at the cheaper rates as compare to the markets rates. It is the difference between operating and maintenance cost of irrigation infrastructure in the state and irrigation charges recovered from farmers. This may work through provisions of public goods such as canals, dams, tube wells etc. which the government constructs and charges low prices or no prices at all (in special cases)for their use from the farmers. It may also be through cheap private irrigation equipment such as pumping sets.

Seed Subsidy:

High yielding seeds can be provided by the government at low prices, and at the future payment options. The research and development activities needed to produce such productive seeds are also undertaken by the government, the expenditure on these is a sort of subsidy granted to the farmers.

Export Subsidy:

This subsidy is given to the farmers to face the international completion. When a farmer or exporter sells agricultural products in foreign market, he earns money for himself, as well as foreign exchange for the country. Therefore, agricultural exports are generally encouraged as long as these do not harm the domestic economy. Subsides provided to encourage exports are referred as export subsidies.

Credit Subsidy:

It is the difference between interest charged from farmers, and actual cost of providing credit, plus other costs such as write-offs bad loans. Availability of credit is a major problem for poor farmers. They do not have sufficient cash to purchase agricultural equipments and cannot approach the credit market because they do not have the collateral needed for loans. To carry out production activities they approach the local money lenders. Taking advantage of the helplessness of the poor farmers the lenders charge very high rates of interest. Many times even the farmers who have some collateral cannot avail loans because banking institutions are mainly urban based and many a times they do not indulge in agricultural credit operations, which is considered to be risky. To tackle these problems the government have provided following provisions:

1. More banking operations in rural areas-which will advance agricultural loans, and
2. The interest rates can be maintained low through subsidization schemes, and
3. The terms of credit (such as collateral requirements) can be relaxed for the poor.

Agriculture Infrastructure subsidy:

Private efforts in many areas do not prove to be sufficient to improve agricultural production. Good roads, storage facilities, power, information about the market, transportation to the ports, etc. are vital for production and sale operations. These facilities are in the domain of public goods, the costs of which are huge and whose benefits accrue to all the cultivators in an area. No individual farmer will come forward to provide these facilities because of their bulkiness and inherent problems related to revenue collections (no one can be excluded from its benefit on the ground of non-payment). Therefore the government takes the responsibility of providing these and given the condition of Indian farmers a lower price can be charged from the poorer farmers.

Chapter 11

ASSIMILATING MARKET INFORMATION

11.1 MARKETING INFORMATION OF CROP

11.1.1 Introduction

Marketing is an important factor in determining the success of any crop or related product. It includes all the operations and decisions made by producer/grower which can be determination of most marketable crops for production, best delivery quality produce etc. However, contrary to popular belief, marketing does not begin after a crop is produced. Instead, marketing alternatives need to be considered even before production takes place.

Marketing of agriculture produce includes assembling, handling, storage, transport, processing, wholesaling, retailing and export of agricultural commodities with supporting services like market information, establishment of grades and standards, commodity trade, financing and price risk management.

- ☆ The marketing system starts with assembling and collecting points located in rural areas close to the growers/producers.
- ☆ The next stage involves storage areas at national grains marketing facilities owned and operated by parastatal and/or private grain elevator; and grain milling companies which in many countries are privately owned and in others are government enterprises.

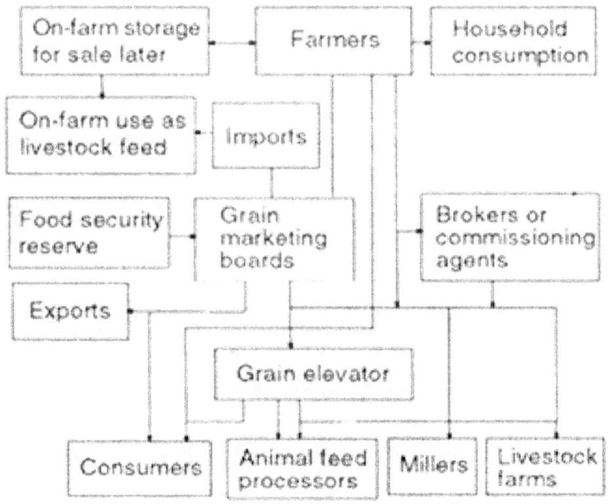

Fig marketing system

- ☆ As size and methods of operation differs from country to country, local assembling and collection points will have grains directly by the farmer-producers themselves or by rural entrepreneurs.
- ☆ In grains, assembly and storage functions are typically combined at marketing stage.
- ☆ In countries where marketing parastatal are in monopoly in grain trading, private traders are sometimes authorised to buy grains from farmers on behalf of parastatal.
- ☆ A common feature of grain marketing systems is co-existence of government marketing agency (parastatal) and parallel private marketing channel with myriad's of private traders.
- ☆ Public grain marketing agencies are government appointed parastatals assigned to control or regulate the system.
- ☆ Earlier than market liberalization, marketing parastatals were recognised in many developing countries who are responsible for setting the prices for major cereals.
- ☆ As per the country, agencies normally consist of one or more ministries related to cereals production and influencing public policies that affects food production and consumption and a parastatal established to operationalise regulatory provisions of public policies enacted for particular crop.
- ☆ In many developing countries parastatals remain important players in grain marketing system even though their role may have changed.

- ☆ In commodity marketing system, private agents include private individuals operating in system as petty assemblers, traders large-scale merchants, millers, brokers and retailers of grain products.
- ☆ The agents in channel operate in rural areas and penetrate remotest areas of rural areas to purchase grains.
- ☆ In developing countries where national structural adjustment programme have not yet been initiated, inter-provincial grain movement controls shows private agents in system cannot legally and profitably transport grain from one production zone to another.
- ☆ For countries that are adjusting their economies structurally, market liberalisation is making or will soon make much restrictions on commodity movement irrational and new form of intermediary are coming into being.

11.1.2 Marketing System of Wheat

In India, the current marketing system for wheat includes private and public system, traders and processors.

The most common marketing channels for wheat are;

Private

- ☆ The Marketing channels in private sector are –
- ☆ Farmer-- Consumer
- ☆ Farmer-- Retailer or village trader-- Consumer
- ☆ Farmer-- Wholesaler-- Retailer-- Consumer
- ☆ Farmer-- Village Trader-- Wholesaler-- Retailer-- Consumer
- ☆ Farmer-- Wholesaler-- Miller-- Retailer-- Consumer

Institutional

It covers the public and co-operative sector agencies. It plays a very important role in the procurement and distribution of wheat. The Food Corporation of India is the main agency for procurement, maintenance of buffer stocks and distribution of wheat. The main institutional marketing channels for wheat are as under;

- ☆ Farmer-- (FCI, State Govt., Co-operatives) Govt. Agency-- Fair Price
- ☆ Shop-- Consumer
- ☆ Farmer-- Co-operative Marketing Society-- Retailer-- Consumer
- ☆ Farmer-- (FCI, State Govt., Co-operatives)-- Private Trader-- Export

☆

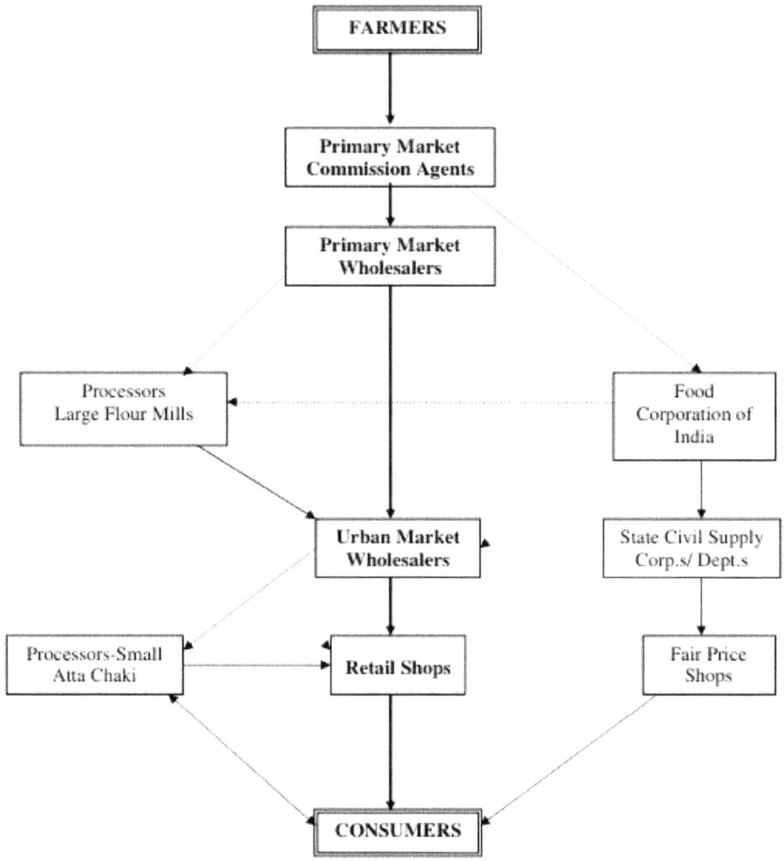

Fig Marketing System of Wheat in India

☆ The system includes Commission on Agricultural Costs and Prices (CACP), Food Corporation of India (FCI), State Civil Supplies Corporations/ Departments.

☆ The CACP studies costs, prices and markets recommends prices to the national government.

☆ The national government's grain management includes procurement, transport, storage and release of grain mainly by FCI.

☆ Grain distribution is done by public distribution system through state governments and fair price shops (ration shops) that spreads across the country in both rural and urban areas.

☆ Over many years, more active minimum support price (MSP) and procurement policies of Government have reduced the role and activities of private traders.

- There are very little private stocking of wheat as the government purchases and stocks large amount of wheat.
- The uncertainty regarding the price and volume of sales from large government stocks often make it difficult for private traders to play their role in market in significant way.
- Many consider the current situation a de facto nationalization of wheat trade.
- The government's purchase and sale policies, together with stock holding and movement restrictions under Essential Commodities Act, state policies, taxation policies, have inhibited private investment in improving grain marketing infrastructure.
- Inefficiency of Food Corporation of India (FCI) exits in terms of handling, transporting and storing of government grain.
- As observed, government is more efficient than private traders, while with FCI setting the base, there might be less incentive for private traders to be more efficient.
- As there appears main role of wheat/rice in farm income and consumer expenditures, inefficiencies in marketing would impose significant costs on producers, consumers, and economy.

11.2 IDENTIFICATION OF SOURCES

11.2.1 Grain Marketing

Marketing systems differ widely as per the commodity, systems of production, culture and traditions of producers and level of development of both particular country and particular sector within that country. The main participants in grain marketing systems are:

- producers,
- marketing boards,
- grain elevators,
- brokers,
- millers,
- farmers,
- animal feed processors,
- millers,

- ☆ food manufacturers,
- ☆ grain exchanges
- ☆ exporters

Wheat growers/producers will better able to use or develop more alternatives if they know the main features of each marketing alternative. Crop marketing alternatives can be direct or non-direct markets.

Direct Market

- ☆ It involves sale of wheat by producer to consumer/miller directly, without any middleman.
- ☆ It allow producers and millers and other bulk buyers to economize transportation cost and improve price realization.
- ☆ Direct market are being run at expenses of State exchequer as a promotional measure to encourage marketing by small and marginal farmers.
- ☆ In this, big companies enters into contracts with producers for buying their produce.
- ☆ Flour millers can directly negotiate with producers and can immediately buy their wheat at farm gates.

Direct market involves producer interaction with consumers on one-on-one basis which include:

- ☆ pick-your-own operations
- ☆ roadside stands
- ☆ community supported agriculture
- ☆ farmers markets

Benefits of Direct Market

- ☆ Increases price realization of the producer.
- ☆ Minimizes marketing cost, transportation cost.
- ☆ Encourages distributional efficiency.
- ☆ Satisfies consumer through better quality of produce at reasonable price.
- ☆ Encourage direct interaction between producers and consumers.
- ☆ Encourage farmers for retail sale of their produce with involvement in marketing process.
- ☆ Help farmers in discovering the demand of markets for future market oriented planning.

Non-direct markets involve producer interaction with market intermediaries which include:

- ☆ terminal market firms
- ☆ shipping point firms
- ☆ processors
- ☆ grower cooperatives
- ☆ brokers
- ☆ retail outlets

11.3 TYPES OF MARKETS:

Markets can be classified based on:

Location: On the basis of place of location or operation, markets can be:

- ☆ Village Markets
- ☆ Primary wholesale Markets
- ☆ Secondary wholesale Markets
- ☆ Terminal Markets
- ☆ Seaboard Markets

Area/Coverage: On the basis of area from where buyers/sellers usually approaches for transactions can be classified as:

- ☆ Local or Village Markets
- ☆ Regional Markets
- ☆ National Markets
- ☆ World Market

Time Span:

- ☆ Short-period Markets
- ☆ Long-period Markets
- ☆ Secular Markets

Volume of Transactions: There are two types of markets based on volume of transactions such as:

- ☆ Wholesale Markets
- ☆ Retail Markets

Nature of Transactions: The markets which are based on types of transactions where people are engaged can be:

- ☆ Spot or Cash Markets
- ☆ Forward Markets

Number of Commodities as per Transactions: A market can be general/specialized based on number of commodities where transactions are completed that includes:

- ☆ General Markets
- ☆ Specialized Markets

Degree of Competition: Market can be placed based on continuous scale, right from perfectly competitive to pure monopoly, so on basis of competition, markets can be:

Perfect Markets: Where following conditions hold true, such as:

a) Large number of buyers and sellers;

b) Buyers and sellers to have perfect knowledge about demand, supply and prices;

c) Prices at any one time are uniform over geographical area, plus/minus cost of supplies from surplus to deficit areas;

d) Prices are uniform at any one place over periods of time, plus/minus cost of storage from one period to another;

e) Prices of different forms of product are uniform, plus/minus cost of converting product from one form to another.

Imperfect Markets: Where conditions of perfect competition are lacking which includes:

a) Monopoly Market

b) Duopoly Market

c) Oligopoly Market

d) Monopolistic competition

Nature of Commodities: Based on type of goods dealt in, markets can be:

- ☆ Commodity Markets
- ☆ Capital Markets

Stage of Marketing: Based on stage of marketing, markets can be:

- ☆ Producing Markets
- ☆ Consuming Markets

Extent of Public Intervention: Based on extent of public intervention, markets can be:

- ☆ Regulated Markets
- ☆ Unregulated Markets

Type of Population Served: Based on type of population served, market can be:

- ☆ Urban Market
- ☆ Rural Market

11.4 SOURCE OF INFORMATION AT VILLAGE AND MARKET LEVEL FOR MARKETING THE PRODUCE

Agriculture information should reach to farmers in right perspective, in right form and in right time as it is very critical for enhancing the productivity of farm sector.

Fig Getting information on telephone

Farmers are getting agricultural information from diverse communication sources like neighbours, local leaders, extension agents, newspapers, telephone, farm magazines, television, radio etc. In context of rural development, nothing is more important than the transfer of useful ideas from one person to another.

11.4.1 Sources of information

As information on agriculture in terms of crop and livestock were communicated among farmers from ancient times, so with development in agricultural research, the need arises to rapidly and accurately transfer new information and technologies to farmers.

To fulfill such need, media such as print and electronic and interpersonal communication has proved its usefulness in creating awareness about new agricultural technologies.

1. Interpersonal sources

It includes:

- ☆ personal localite
- ☆ extension agents
- ☆ friends
- ☆ family members
- ☆ fellow farmers
- ☆ neighbours
- ☆ group discussions
- ☆ demonstrations
- ☆ campaign
- ☆ village leaders

The sources play significant role in dissemination of information related to agriculture. During pre-independence, interpersonal communication channels were used widely. Demonstration centers were established to disseminate information by conducted demonstration under guidance of experts.

2. Mass media

- ☆ Mass media serves as channels of communication that expose large numbers of people to similar information at same time.
- ☆ It includes:
    - media that conveys information by sound such as radio, audio cassettes;
    - moving pictures such as television, film, video;
    - print such as posters, newspapers, leaflets
- ☆ The attraction of mass media to extension services is its high speed with low cost through which information can be communicated to people over wide area.
- ☆ As cost of producing and transmitting a radio programme may seem high, when that cost is divided between millions of people who may hear programme, it results as cheaper way of information.
- ☆ The cost of an hour's radio broadcast per farmer who listens can be less than one-hundredth of cost of an hour's contact with an extension agent.

3. Radio

- ☆ It is a popular medium for information as well as attitude change.
- ☆ Radio from its inception played an important role in development communication.
- ☆ It plays a peculiar role in technology dissemination.
- ☆ It is very effective media which can reach effectively to rural people.
- ☆ In India, All India Radio has been the forerunner in process of implementing communication strategy adopted by government.
- ☆ Radio communication works roughly in 15000 villages and carries certain programs like Utam Khetti, Kisan Vani, Navi Chetana on agriculture.

4. Television

- ☆ It is an ideal medium to convey information and news to illiterate and literate, urban and rural viewers on whom it would have profound impact.
- ☆ It is used in planned manner to motivate people to participate in developmental programs.
- ☆ It is introduced in India in 1959 on experimental basis.
- ☆ In 1967, Delhi Television centre launched Krishi Darshan Program which gained popularity as they have been able to solve numerous ground level problems of farmers and also replaced their antiquated technique with new innovations.

11.4.2 Rural Marketing

Marketing in rural involves development, pricing, promotion, distribution of rural specific product and services which leads to exchange between rural and urban market satisfying consumer demand and obtaining organizational objectives.

It can be:

Urban to Rural:

- ☆ It involves selling of products and services by urban marketers in rural areas.

Rural to Urban:

- ☆ In this, rural producer sells the product in urban market.
- ☆ It is not direct.
- ☆ It is carried out using middlemen, agencies, and government co-operatives.

Rural to rural:

- ☆ It includes selling of agricultural products to other village in its proximity.

11.5 EFFECTIVENESS OF MASS MEDIA/INTERPERSONAL COMMUNICATION SOURCES AS PERCEIVED BY WHEAT GROWERS

- ☆ As studied on farmer's information source and their relative effectiveness, it is noticed that print media were regarded as most effective source, as they were used as source of information by only 11% of respondents.
- ☆ The information sources for agricultural technology found that contact farmers were accorded first rank by respondents on credibility.
- ☆ The role of mass media in dissemination of agricultural technologies among farmers revealed that T.V was ranked 1st with regard to its effectiveness in dissemination of agricultural technologies while newspapers, posters, charts, books/booklets and bulletins ranked 3rd, 4th, 5th, 6th and 7th on basis of effectiveness.
- ☆ With role of television in agriculture development, majority of respondents felt that television is indeed the best medium of disseminating agricultural related information to general public.
- ☆ Many farmers found that farm magazine gives highest satisfaction followed by telephone, television, computer and radio.

11.6 PRECAUTIONS BEFORE MARKETING THE PRODUCE

Before marketing of wheat produce, wheat grower should monitor the wheat produce in terms of dryness, grading, bagging. Wheat which are not to be marketed for current year, should be properly stored.

11.6.1 Proper drying of wheat for marketing

Wheat grower before market the wheat produce, should properly check for dryness of wheat. The idea of wheat drying is to reduce grain moisture content so that it should meet the required levels for safe, long-term usage/storage.

If the wheat is not to be marketed at current time, should be correctly stored. Wheat produce if placed in storage should be dried quickly to moisture level of about 12% in order to minimize any quality deterioration. Drying of wheat can be done in bins by blowing large amount of dry air through the grain.

It is very important to maintain on-farm wheat drying cost at minimum in order to maximize profits. In order to have better wheat crop/grain quality, wheat producers/ grower should follow following steps, which will help in proper monitoring and checking of wheat moisture level:

- ☆ Wheat should be harvested at 20% or less moisture content as it needs extra care, if harvested above 20% moisture content.

- ☆ Load wheat grains in cleaned and sanitized bins after harvesting.
- ☆ Check moisture content of each load of grain as it is placed in drying bin.
- ☆ Open air exits and start the fan as soon as the grain depth is about 1 foot deep on perforated floor.
- ☆ Be sure to use spreading devices or some other means to keep the grain leveled as the bin is being filled.
- ☆ If the grain is allowed to cone, there will be an increase of small particles in center of cone/ or central portion of bin.
- ☆ Add wheat to drying bin in shallow layers till moisture content decreases.
- ☆ High moisture wheat (18 to 20%) can be added in 4 feet layers on top of dry grain if fan can provide at least 3 to 4 cfm/bu through the total depth in the bin.
- ☆ Level wheat inside each drying bin continuously and never allow coning to take place.
- ☆ Use stirring devices when drying wheat if possible as temperature can be set as high as 130°F.
- ☆ Monitor the moisture content of wheat daily.
- ☆ If the heat has been on long enough for complete mass of wheat to be warmed and the weather is clear and dry with humidity below 60%, turn the heat off when the moisture content of the grain drops to within 1% of the target moisture content.
- ☆ Continue running the fans, and the residual heat in the grain will finish the drying process.
- ☆ Aerate wheat with natural air once the grain is below 13% moisture content.
- ☆ Wheat should be cooled as much as possible with early summer conditions.
- ☆ Cooling air should be checked for humidity, being careful to aerate when humidity is below about 60% or better yet when the EMC content is at or below the target moisture level.
- ☆ Avoid aeration with high humidity air since it will add moisture back to the grain.
- ☆ Probe the bin periodically to check for insect infestation and grain temperature increase.
- ☆ Wheat temperature increase usually means moisture migration.

11.6.2 Grading of wheat for marketing

Market quality is identified if grade is established for wheat crop. One of the main objectives of grading is to ensure uniform quality, so that wheat produce can be priced to satisfy the interests of producer and consumer. The grading standards employed and ability of handling or distributing to maintain standards till the produce is delivered to the customer.

Grain grading and specification systems for grains ensure that parties involved in production, storage and processing of grain can handle, trade and process grain which complies with required standards or properties.

In many countries, grading of grain depends on following properties:

1. Hectolitre mass
2. moisture content
3. foreign matter
4. damaged grains

Benefits in Marketing

Grading of wheat produce will give following marketing benefits:

- ☆ Low expenditure in transport and storage.
- ☆ Knowledge of prevailing price, and right markets.
- ☆ Easy financial assistance, and future trading.
- ☆ Widens the market for farm products.
- ☆ Consumers get wide choice of quality at a reasonable price.
- ☆ Promotes competitive marketing.

Grade Specifications:

Grading is undertaken by different agencies as per different criterias on usage. The bulk of commodity may be simply classified based on kernel texture such as:

1. hard
2. semi hard
3. soft with colours (white, amber and red)

Factors constituting grain quality

Factors which constitute quality are:

1. impurities, which together with damaged grain may seriously affect milling quality
2. bushel and kernel weight

3. nature and structure of kernel
4. moisture content

The merchants consider mainly physical characters like impurity or refraction and general appearance of the grain.

Agmark Grade Characteristics:

- wheat shall be dried grains of Triticum vulgare and Triticum durum Desf;
- have uniform size, shape and colour;
- sweet, hard, clean, whole-some and free from moulds, weevils obnoxious smell, discolouration, admixture of deleterious substances and all other impurities except to the extent indicated in the Schedule;
- sound merchantable condition;
- not having moisture exceeding 12%

Advantages of Grading

- There are many advantages of grading to different groups of persons which includes:

1. Grading before sale allow farmers to get higher price for their produce. It serves as an incentive to producers to market the produce of better quality.
2. Grading facilitates marketing, for size, color, qualities and other grade designations of product that are well known to parties and requires no assurance from seller about product quality.
3. Grading widens the market for the product, for buying which can be between the parties at distant places on telephone without any inspection of product quality.
4. Grading reduces the cost of marketing by minimizing expenses on physical inspection of produce, thereby minimizing storage loses.
5. Grading minimizing advertisement expenses and eliminating the cost of handling and weighing at ever stage.
6. Grading makes it possible for farmers to get easy finance when commodities are stored, getting claims by railways/insurance companies, storage place for produce, market information, produce of different farmers, facilitate futures trading in commodity.
7. Grading helps consumers to get standard quality products at fair prices.
8. Grading contributes to market competition and pricing efficiency

11.6.3 Bagging and transportation of wheat for marketing

Bagging

Fig Bags of grain

Proper packaging is an important element in reducing losses, especially in the tropics, climate considerably increases the risks of grain deterioration. The main functions of packaging are to:

- ☆ permit easy handling, whether manual or mechanical;
- ☆ reduce product losses by theft;
- ☆ protect product from external attack (humidity, insects, sunlight, etc.).

There are various types of packaging for agricultural commodities, appropriate to the product's characteristics and to the marketing system. The choice of the type of bag should take into account not only its inherent toughness and resistance to humidity, sunlight and pests but also the type of handling anticipated.

Wheat grains are normally stored in 40-80 kg bags which are made either from jute or woven plastic. As per the size of storage, these bags are normally formed into stack. When using bag storage, consideration needs to be given to following:

- ☆ Jute bags should not be stacked higher than 4 m and plastic bags 3 m as plastic bags are more slippery and stacks will be less stable.
- ☆ Bags should be stacked under cover e.g. under a roof, in a shed or granary or under water proof tarpaulins.
- ☆ A one meter gap should be left between and around stacks and 1.5 m clearance between the top of the stack and the roof.
- ☆ Bags should be stacked on pallets or on an above ground structure to avoid the possibility of absorbing moisture from the floor.
- ☆ Bags should not be stacked on a bed of wheat husks or bags filled with wheat husks, as these are difficult to keep free from insect infestation.

☆ Bags should be stacked so that fumigation can be undertaken easily. The dimensions of the stacks should be set to facilitate sealing with a single fumigation sheet.

☆ The efficiency of bag storage can be improved if hermetic plastic liner bag is used inside the existing storage bag especially for seeds but also for commerical wheat.

Storing of wheat grain bag

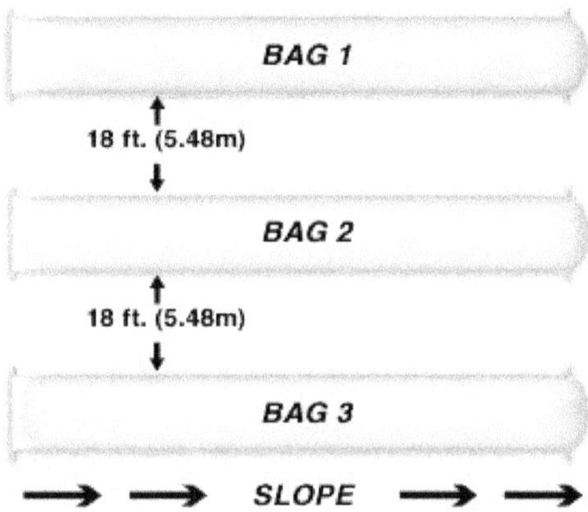

☆ The wheat grain bags should be placed in north - south direction on storage site as it allow grain bags to be exposed to sun evenly.

☆ If wheat grain bags are positioned in east - west direction, then south side of grain bags will be exposed to sun and may lead to overstretching and damage to grain bags.

☆ Keep the grain bag on highest elevation of storage site as it will aid in loading of grain bag and assist in preventing any moisture from entering the bag at the start and at finish points.

☆ Do not keep the wheat grain bag across the slope as it could cause water build-up along the edge of bag which will add stress on one side of bag and may led to overstretch of bag.

☆ While storing wheat grain bags side by side, it should be kept in mind that there should be proper space of around 18 ft. between two bags, as it will help for unloading equipment and trucks to travel between the bags.

☆ If the grain bags are stored in the field, keep in mind that bags should be positioned in a line as it will help in preventing from animals, rodents and other pests.

☆ If the wheat grain bags are stored in he depots, then position the bags in pairs with 5 ft. distance apart or 18 ft. in case of pairs.

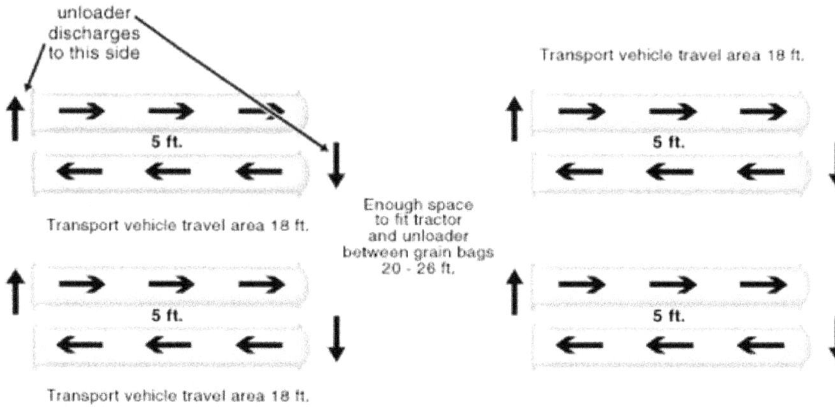

Precautions

For well-managed reception and storage of wheat grain delivered or stored in bags, following rules should be observed:

☆ avoid taking in more grain than storehouse can hold as capacity depends on number of individual lots one wishes to store;

☆ store only grains that are very dry and clean;

☆ recondition products in case of damp or torn bags, or if the quality of grain seems doubtful;

☆ erect stacks of bags that are stable and easily accessible;

☆ maintain the cleanliness and good condition of premises and surroundings, equipment, and stored products;

☆ make timely provision for supplies of fuel, bags, insecticides and miscellaneous products.

Sealing of Grain Bag for Marketing

- ☆ When the grain bag is full, seal the wheat grain bag as soon as possible so as to remove the chance of excess moisture that enters the bag and can damage the grains.

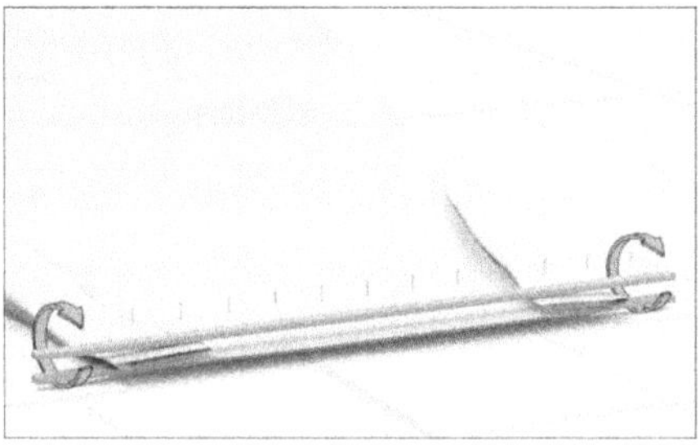

Fig Sealing of grain bag

Sealing of Grain Bag:

- ☆ Remove around 10-13 ft. of grain bag from the loader.
- ☆ Align the bag ends and fold the two outer edges of the bag.
- ☆ Start the fold approximately 2 ft. up on the bag and fold it towards the center of bag.
- ☆ Place a board underneath the folded end of the bag and keep a second board on top of bag.
- ☆ With folded end of grain bag, place it between the two boards and fasten the boards together.
- ☆ Flatten the end of bag, roll the sealed end downward three times and fold it underneath the grain bag.
- ☆ As grain bag is being filled, weight of grain will provide downward pressure on seal and help create an additional seal to help preventing moisture from entering the grain bag.

Transport

Transportation involves movement of crop/cereals between the places which acts as one of the important marketing functions at all stage. The main advantages of transportation of goods are:

- ☆ Widening of Market
- ☆ Narrowing Price Difference Over Space
- ☆ Creation of Employment
- ☆ Facilitation of Specialized Farming
- ☆ Transformation of Economy
- ☆ Mobility of factors of Production

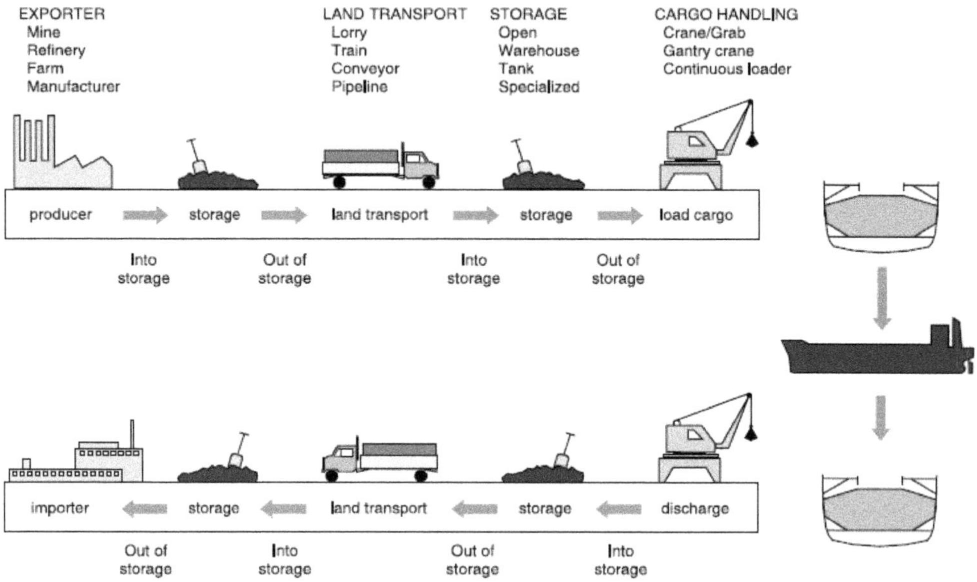

- ☆ Labor intensive systems of grain movement serve to minimize capital investment in India where cost of labor is low.
- ☆ Wheat bags are loaded and unloaded by manually/machine from wagons, trucks, railroad cars, and barges between farm and mill.

- ☆ In some situations, bagged wheat may be loaded on and off the vehicles many times manually before it is milled.

- In post-harvest handling, transport and storage of grains at farm level is done partially in bulk.
- The transportation of grain to primary markets by farmers is done in bulk using bullock carts, tractor trolleys or lorries.

- At the market yard, grain is displayed in bulk, auctioned, cleaned, bagged, weighed and delivered to consumers in bags.
- The food grain trade depends upon labor, so handling, transport and storage of marketed grains in bags is common.
- Problems arise when old torn bags are used which spill grain, causing loss.
- Mostly 100-kg bags are used which are cumbersome to carry.
- Other hazards for bags are hooks which tear the bags, the rough surface of the carts and trolleys and nails, which damage sacks when they are pulled.

- Transportation occurs from farm to market, market to consumer, market to temporary storage, temporary storage to long term storage and long term storage to consumers.
- Since wheat is transported in bulk and bags from field to market, so the following means of transportation are used at different stages of marketing which includes:

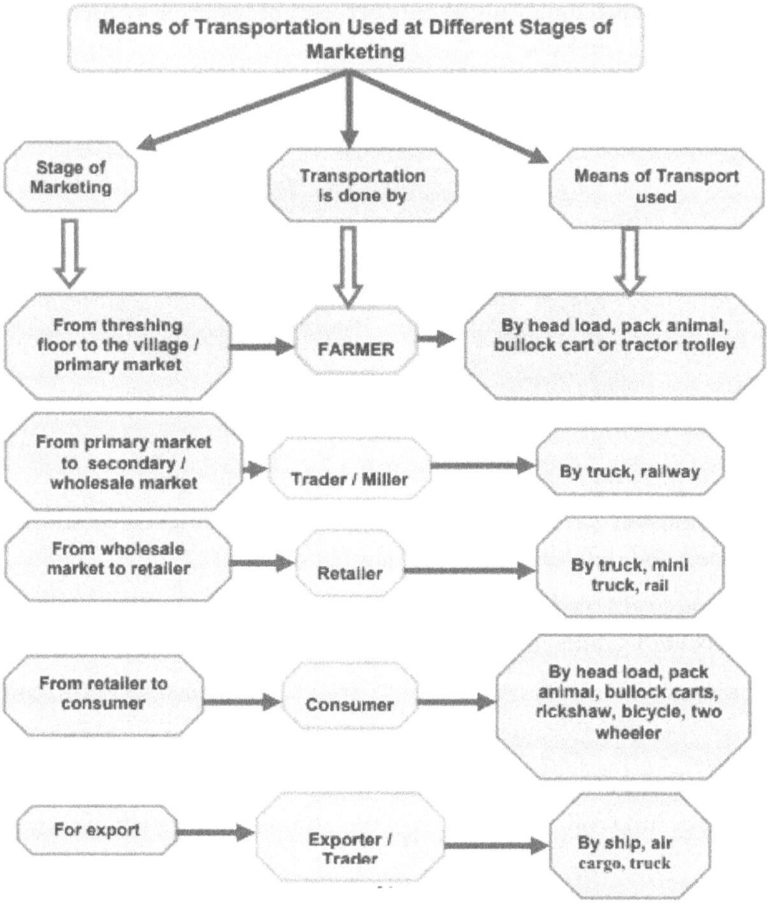

Fig Means of transportation of wheat

Roads:

- ☆ The road transport is generally used for transporting of wheat crop in internal markets.
- ☆ It is also used for transporting wheat crops to neighbouring countries that are connected by roads.

Railways:

- ☆ This is most important and cheapest mode of transport for wheat.
- ☆ It is good for long distance and for transporting larger quantities.
- ☆ It needs more handling charges as it requires loading and unloading and local transport cost.

Water/Sea transport:

- ☆ Transporting wheat/crops from sea is an oldest and cheapest modes of transport.
- ☆ This transport is a little slow but cheap and useful for carrying large quantity.
- ☆ Export of wheat is done mostly by ship through sea.

11.7 INFORMATION ABOUT PRICE FLUCTUATION IN MARKET

11.7.1 Market Price

Movement of products from producers to consumers involves costs, taxes and cess which together contribute to market price. Such price vary with channels by which a particular crop/commodity passes through such as cost of packing, transport, weighment, loading, unloading, losses and spoilages.

Market price normally includes:

- ☆ Handling charges at local point
- ☆ Assembling charges

- Transportation and storage charges
- Handling by wholesaler and retailer charges to customers
- Expenses on secondary service like financing, risk taking and market intelligence
- Profit margins taken out by different agencies.
- Producer's share in consumer's rupee

Wheat grower/distributor should know the actual market price of their product which will help them in:

- ascertaining which intermediaries are involved between producer and consumer.
- ascertaining the total cost of marketing process of commodity.
- comparing the price paid by consumer with price received by producer.
- seeing whether there is any alternative to reduce the cost of marketing.

11.7.2 Price fluctuation in market

Wheat/Grain price volatility has remained a major concern despite considerable advances in communication, transportation and production technologies as grain production is closely associated with national food security and welfare of farming population, so managing price instability has been a major challenge. Wheat/grain producers want prices to be stabilized at levels that guarantee adequate returns but interventions to achieve such objective are often costly to consumers and tax payers.

The volatility is often attributed to special characteristics of agriculture which includes:

- supply disturbances (climate and pests)
- inelastic supply and demand
- production decisions based on current prices
- fluctuating food
- agricultural policies

Price volatility that results from peaks and troughs in production is often made worse by host of endogenous causes such as:

- lack of adequate storage facilities
- low stock levels
- sudden jumps in input prices

- ☆ underdeveloped market infrastructure
- ☆ limited export and import

High levels of production and low prices in agricultural sector also follow shortages and high prices in unstable cobweb process if production decisions at planting time are based on expectation that current prices will remain in place after harvest. If current prices are high, wheat growers/producers assume that this situation will continue and will accordingly take measures to expand production for next season, which results in high levels of production and price collapse.

There are three main reasons why wheat/commodity prices move higher or lower:

- ☆ Fundamental state of commodity market: If current inventories exceed demand, then oversupply tends to drive prices lower, but if the demand is more than supply, then inventory deficit tends to push prices higher.
- ☆ Technical condition of market: Price in wheat/commodity fluctuates due to market technical conditions. Price charts often drive the behavior of investors, traders and other market participants, so as everyone studies same data, a herd mentality of massive group buying or selling consequently influence prices.
- ☆ Price is sensitive to changes in global macroeconomic and geopolitical landscape.

11.7.3 Decision based on Price Fluctuation

Wheat growers/farmers mostly have downside price risk in contrast to end user who mostly has upside price risk. Managing price risk can occur through a spot transaction if the product is available or by forward transaction if crop/product is not available, either with physical forward contract or hedging price by futures market.

Selecting subjective price targets/triggers depends on individual risk attitude and circumstance of decision maker. An unrealistic target price will add profit margin to current price in bearish market when commodity market intrinsically ignores operational costs of farmer or end user. Farmers may sometimes be forced to accept prices below their cost of production, whereas end users may sometimes be forced to buy above their break-even limit. The alternative for farmers is to withhold product from the supply chain either by hoarding or pooling, whereas the alternative for end users is a switch to substitute products or alternate operating methods.

To minimize farmers/wheat growers price risk, government has started with Minimum Support Price (MSP) which is a form of market intervention that insure agricultural producers against any sharp fall in crop/cereal prices. The minimum support prices are announced by government at the start of crop sowing season on the basis of recommendations of Commission for Agricultural Costs and Prices (CACP). MSP is price fixed by government to protect the wheat growers/producer/ farmers against excessive fall in price during bumper production years.

The minimum support prices are guarantee price for crop produce from the government. Its main objectives is to support the farmers from distress sales and to procure food grains for public distribution. In case the market price for commodity falls below the announced minimum price due to bumper production and glut in the market, government agencies purchase the entire quantity offered by the farmers at the announced minimum price.

Chapter 12

MAINTAIN HEALTH AND SAFETY AT WORKPLACE

12.1 BASIC SAFETY CHECKS BEFORE OPERATING VEHICLES AND ALL OTHER MACHINERY

12.1.1 Injuries due to Farm machinery/vehicle

Farm equipments and vehicles daily check are simple and effective way to spot potentially dangerous issues or defects before they are used. Farmers or operators should need to have a system of daily routine checks for machineries and equipments that are used in farms to ensure that are in good working order, safe and fit for purpose before start of working day or work shift.

Farmers and farm workers use tractors and other equipment every day. These machines are powerful, and if not operated safely, can lead to disabling injuries or death. Modern tractors, telehandlers and other farm equipment have given farmers with an ability to work more quickly and efficiently, but these machinery brings new dangers.

These agricultural machines have potentially dangerous moving parts, which can cause serious or fatal injuries such as:

- ☆ balers – pick-ups, twine mechanisms and moving rear doors;
- ☆ forage harvesters – chopping cylinders;

- ☆ combine harvesters – augers in the grain tank and the header unit;
- ☆ potato harvesters – rotating rollers and conveyors;
- ☆ slurry tankers – power take-off (PTO) shafts;
- ☆ bale and straw choppers – chopping mechanisms;
- ☆ tractor – hitch mechanisms, PTOs and PTO shafts;
- ☆ power harrows – rotating tines;
- ☆ feeder wagons – rotating components in the mixing chamber.

Poorly used or faulty machinery is the main cause of death and injury on farms. Machinery accidents can be prevented by keeping the machine in good repair, fitting and ensuring all safety equipment like guards, safe access platforms and ROPS to be operational with machine all the times.

Farmers and farm workers before using the machines or vehicles should ask themselves:

- ☆ Is the machine they intend to use is suitable for the job?
- ☆ Are all safety devices like guards are in place and are working correctly?
- ☆ Are machine operator properly trained to do this job/use this machine?
- ☆ Has the instruction manual for the machine been provided, read and understood?
- ☆ Is the right personal protective equipment is available and worn?
- ☆ Has a risk assessment been carried out?
- ☆ Has the work been properly planned and communicated to those who may be at risk?
- ☆ If two or more people are involved, has everyone understood what needs to be done and has a system of communication been agreed on?
- ☆ Are all operators/workers trained to do this job?
- ☆ Is suitable clothing and footwear available and worn (including any PPE)?
- ☆ Have operators been provided with relevant information?

12.1.2 Checking the machine before use

Before working with any machinery, farmer/operator should carry out the basic check to make sure that the machine/vehicle what he/she will be using is in the good working order and is safe for use. For this, the machine/vehicle should be checked with basic checking like:

- ☆ Mechanical defects – pay particular attention to items like brakes, wheels and tyres.
- ☆ Guards and other protective devices should be correctly fitted and maintained in good condition.
- ☆ Check that Guards and other protective devices are in correct place, securely attached and working properly.
- ☆ The damaged or defective guards/devices should be repaired or replaced before the machine is used.
- ☆ Check that stopping devices are functioning correctly, such as emergency stops.
- ☆ See that all controls should be clearly marked to show what they do.
- ☆ If workers are to be carried on the machine ensure that they can do so safely, eg check for safe means of access and that working platforms have guard rails etc.
- ☆ Hitching and attachment points – check that mounted, semi-mounted or trailed machinery are safely attached to the towing vehicle such as a tractor.
- ☆ Examine the conditions of drawbar/pick-up hitch, hitch rings, pins, clips etc.
- ☆ Carry out any pre-use checks as specified in the operator's manual.
- ☆ For self-propelled machines, make sure that mirrors are clean and properly adjusted.
- ☆ Check any other reversing aids are working.

Checking Grain Harvesting Equipment: Safety before Starting

Farmer/operator should check for following before using grain harvesting equipment:

- ☆ Before trying to operate a combine, operator/farmer should study the operator's manual which carries information about all general safety rules with specific safety recommendations for particular machine.
- ☆ Farmer/operator if checking the combine inside the building should make sure that all doors are opened to provide good ventilation as exhaust fumes from gasoline or diesel engine are poisonous which can lead to series health hazard.
- ☆ Always clean the combine before starting as trash around the exhaust system can cause fires.
- ☆ Clean the oil, grease or mud on ladders or on platform of the combine as they can cause serious falls.

- ☆ Check if combine is equipped with cab, then clean the glass to provide maximum visibility.
- ☆ Check the tire pressure of combine every day as under-inflation can lead to buckling of sidewall, which causes dangerous tire failure.
- ☆ Check the brakes once in a week as with hydraulic brakes, it is required that the master cylinder should have full fluid and should be free from air inside the lines.
- ☆ Check the threshing cylinder-rocking bar to see it is clear of the cylinder.
- ☆ Make sure that all shields and covers are in place and fastened securely.
- ☆ Remove or stow all service equipment.
- ☆ Always use the handrails and ladders provided on the combine for safe mounting and dismounting.
- ☆ Be sure that all PTO covers, safety stands and shields are on the machine before taking it to the field.

Checking Bailing Hay Equipment:

Farmer/operator should check for following before using the bailing hay equipment:

- ☆ Check that the baler should be properly connected to the tractor.
- ☆ Bailer should have required safety guards fitted.
- ☆ See that nobody should be allowed to ride on the baler.
- ☆ See that nobody is getting too close to the baler.
- ☆ Check that fire extinguisher should be fitted to the machine.
- ☆ Disengage PTO and apply fly wheel brake before doing any baler adjustments.
- ☆ Stop engine and apply fly wheel brake before repairs or stringing up the baler.

12.2 APPROPRIATE PROTECTIVE CLOTH AND EQUIPMENTS AT WORKPLACE

12.2.1 Protective Clothing/equipments

Farmers or farm workers are involved in different process such as cutting, harvesting, equipment and vehicle maintenance, so they should use appropriate cloths, tools and equipment to work. As farmers or farm workers work in such areas that are not designed for proper working due to various dangers and hazards, so they are advised to use proper clothing and protective equipment while working at farms.

Clothing is a tool that protects the body while working on farms. Protective clothing is manufactured using traditional textile manufacturing technologies like weaving, knitting and non-wovens and by specialized techniques such as 3D weaving and braiding by natural and man-made fibers. Wearing of proper clothes at work will help farmers or farm workers to prevent from various injuries.

Protective Equipment refers to as any specialized equipment or clothing that are used by farmers and farm workers for protection against health and safety hazards, so they are designed to protect many parts of the body such as:

- ☆ eyes,
- ☆ head,
- ☆ face,
- ☆ hands,
- ☆ feet,
- ☆ ears,
- ☆ torso

These protective equipment not only prevent accidents, but also prevent/reduce injuries and fatalities.

With respect to equipment and tools to be used at site, farmers or farm workers should ensure that:

- ☆ right tool and equipment are available for job
- ☆ tools are in appropriate condition
- ☆ tools are suitable for working environment
- ☆ tool has ergonomic design

Farmers or farm workers should see that all clothing and protective equipment must:

- ☆ be appropriate for risks involved, without itself leading to any increased risk
- ☆ correspond to existing conditions at workplace
- ☆ take account of ergonomic requirements and workers state of health
- ☆ fit the wearer correctly after any necessary adjustment

Since farmers or farm workers are exposed to pesticides and organic dust which serves as most important occupational risks, so to overcome the occupational health hazard, protective clothing/accessories are necessary.

Farmers or farm workers should note that protective clothing should be designed in such a way that they are suitable and acceptable when its functional features provides protection to the wearer without causing any health problem or hindrance while working. The designed protective clothing for farm workers reduces occupational health hazards and increases work efficiency.

12.2.2 Types of Protective Equipment

1. Head Protection:

- ☆ Safety helmets and hard hats can protect a worker from head injuries caused by fire, electrical shocks or falling objects.
- ☆ The helmets used should have hard outer shell to block penetration by blunt or sharp objects.
- ☆ It should also be fire and water resistant and should be able to absorb the shock when struck.

2. Protective Clothing:

- ☆ Wearing protective clothes will save the body from cuts, radiation, dangerous chemicals, hot liquids, contamination etc.
- ☆ The materials of protective clothing varies as per the application and working condition.

3. Face and Eye Protection:

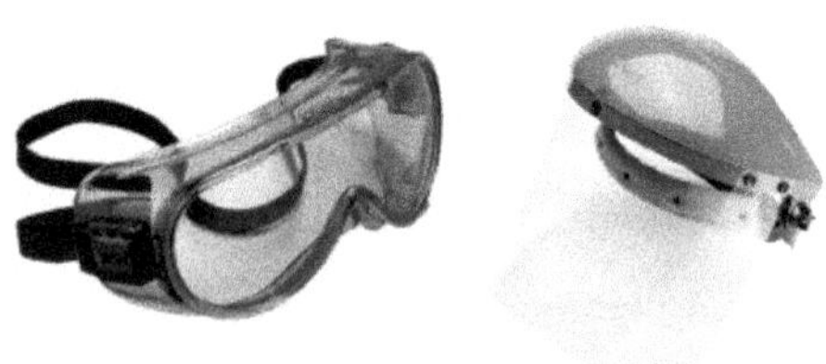

- ☆ They protect the face and eyes from liquid chemicals, gases, UV light, flying debris and molten metal splashes.
- ☆ The floating or flying dust, sparks, swinging ropes, chains and laser light could also injure operators.
- ☆ Se of goggles, protective eyeglasses, welding shields, laser safety goggles and face shields are advised for farm workers.

4. Respiratory Protection:

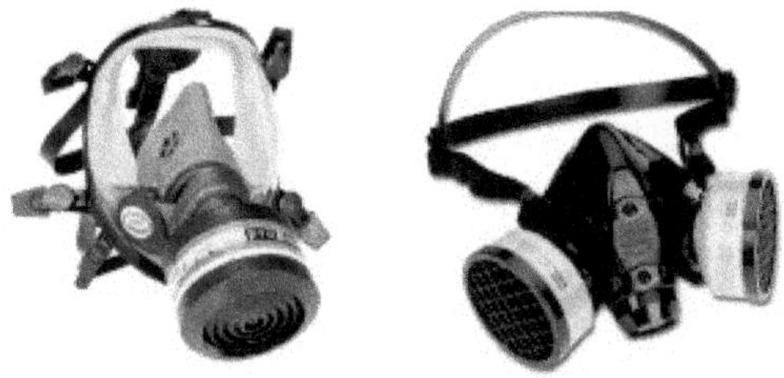

- ☆ Such type of equipments is designed to protect a farm worker from inhaling contaminated air or toxic compounds in workplace.
- ☆ Breathing apparatus, chemical cartridge respirators, powered air purifying respirators and dust masks should be used at workplace.

5. Hand and Arm Protection:

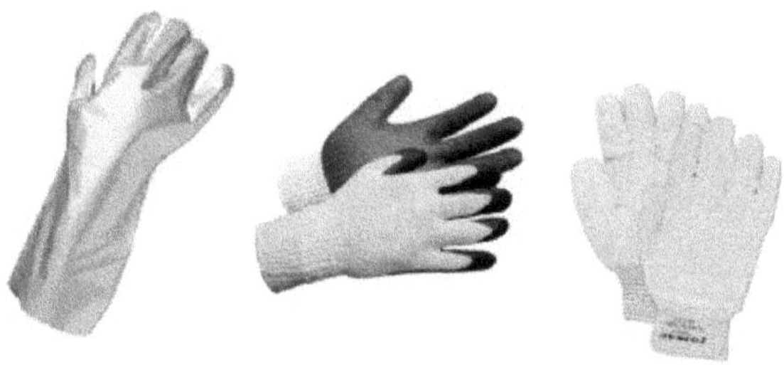

- ☆ Such type of protective equipments is designed to protect the arms and hands from cuts, burns, bruises, abrasions and punctures.
- ☆ Many such protective equipments are designed to protect the farmers/farm workers from heat/ temperature extremes and helps them from preventing amputation and fractures.
- ☆ Gloves (hand or elbow), arm coverings and finger guards should be used by farmers/farm workers while working at workplace.

6. Hearing Protection:

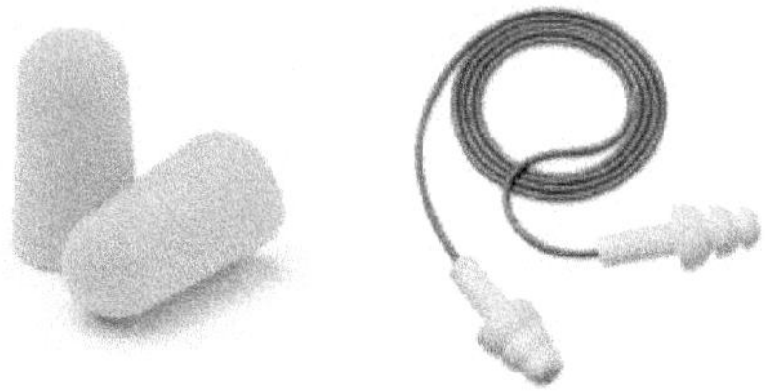

- ☆ Such equipments are designed to protect farm workers from deafening and prolonged noises in the workplace.
- ☆ It includes earplugs and earmuffs.

7. Foot Protection:

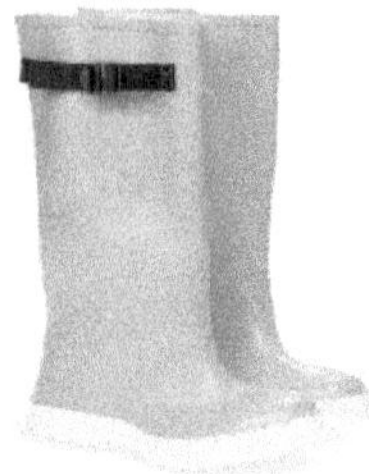

- ☆ Protective footwear must be worn by farm workers while working at workplace, as these will help from risk of falling objects, cuts, soil creatures.

12.2.3 Using Correct Protective Clothing and Tools

Farmers or farm workers are exposed to pesticide vapours and organic dust during various farm activities which may leads to many health problems, so to minimize health hazards, protective clothes were designed with reference to fabric and functional features.

Wearing the right protective cloths will help farm workers from various potential contamination. Correct protective clothing for farm workers include disposable coveralls, chemical-resistant suits, chemical-resistant aprons, and chemical-resistant headgear.

As pesticides are commonly used in farming against pests, diseases and weeds, so depending on the toxicity and usage, they pose health risk if farmers/farm workers do not wear adequate protective clothing. Selecting personal protective equipment for handling hazardous chemicals like pesticides is a complex process and should be made part of workplace overall safety.

Protective clothing for pesticide applicators

- ☆ Dress Jacket with hood
- ☆ Pyjama of water proof fabric (lining of cotton hosiery fabric)
- ☆ Mask resistant to chemicals (Respirator particulate)
- ☆ Glasses/goggles
- ☆ Sports shoes
- ☆ Surgical gloves

Protective clothing for threshing period

- ☆ Apron with hood and full sleeves with elasticized cuffs (for males)
- ☆ Kameez with full sleeves preferably elasticized cuffs (for females)

- ☆ Pleated mask / beak mask
- ☆ Glasses/goggles
- ☆ Sports shoes

Precautions

Farmers or farm workers should:

- ☆ Wear hard hat when doing work related too trimming trees, repairing machinery etc.
- ☆ Wear bump cap for doing everyday farm work.
- ☆ Wear impact-resistant safety eyewear when doing dust-raising or potentially eye damaging types tasks.
- ☆ Use acoustic ear muffs/inserts when using noisy farm machinery and power tools.
- ☆ Use dust mask/filter respirator when working in dust, spray painting, or handling moldy forage.
- ☆ Use gas mask/chemical cartridge respirator when applying pesticides or ammonia fertilizers.
- ☆ Wear self-contained breathing device in any place where there could be toxic fumes or insufficient oxygen.
- ☆ Protect hands by wearing gloves or applying barrier cream; however, don't wear gloves around machinery moving parts.
- ☆ Protect feet with safety shoes or boots with slip-resistant soles.
- ☆ Wear impervious clothing when using toxic or irritating chemicals to protect skin.

12.3 UNDERSTANDING HEALTH HAZARDS BY USE OF HERBICIDES/ PESTICIDES/FUMIGANTS ETC

12.3.1 Introduction

Pesticides

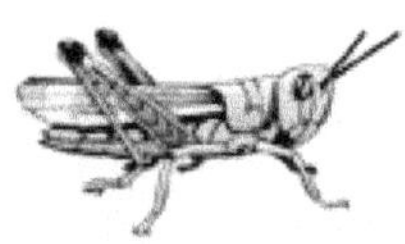

Pesticide is a group which contains:

- ☆ insecticides,
- ☆ fungicides,
- ☆ herbicides,
- ☆ rodenticides,
- ☆ garden chemicals,
- ☆ wood preservatives
- ☆ household disinfectants

Pesticides have been an essential part of agriculture as they protect crops and livestock from pest infestations and gives reduction for many decades. It is poisons and can be hazardous.

Farmers and farm workers are exposed to pesticides in many different situations such as in:

- ☆ Mixing
- ☆ Loading
- ☆ Spraying
- ☆ Storing
- ☆ Spillage
- ☆ Disposal

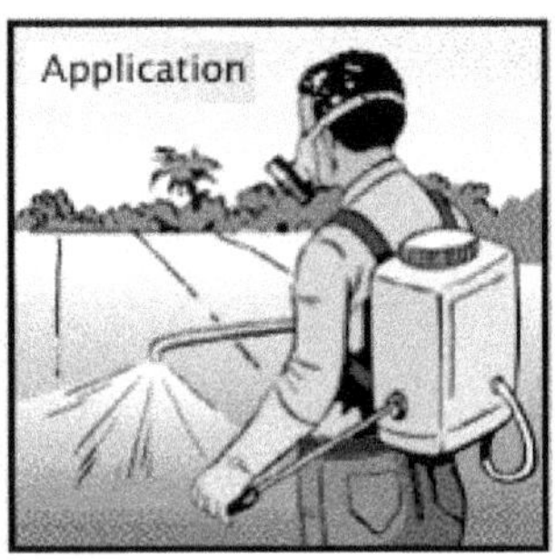

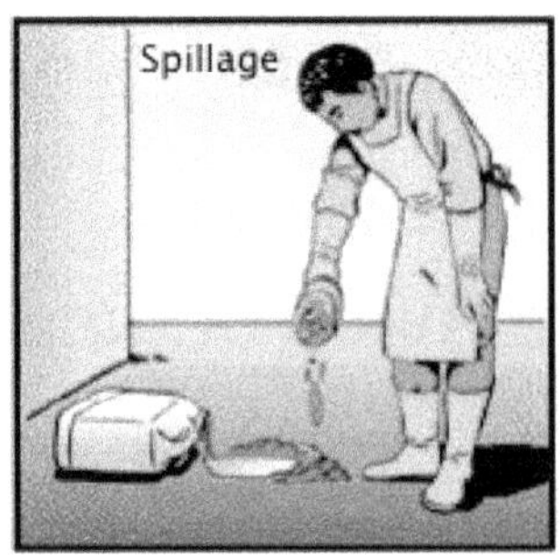

Pesticides enter in body through:

- ☆ Skin
- ☆ Breathing
- ☆ Mouth

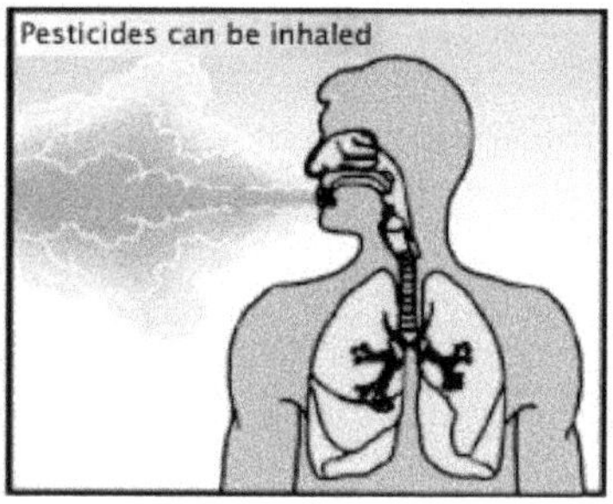

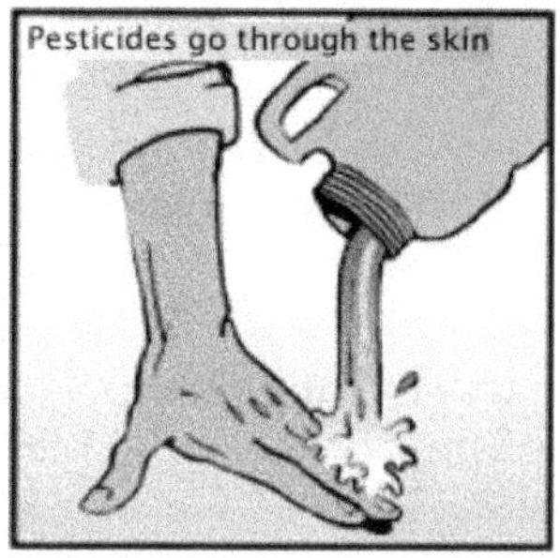

Fumigants

- ☆ Fumigation is pest control method where pesticide gas or vapor is released in air or injected in the soil to kill or eliminate pests.
- ☆ It fills the airspace in a structure with toxic gas.
- ☆ It uses tarp or tent to trap the gas in the area which requires treatment.
- ☆ It is commonly used in agricultural practice to treat fields to reduce weeds, nematodes and other soil insects and small rodents such as moles.
- ☆ In this, the fumigants are injected in soil and move through the soil air which dissolve in the water in the soil where they kill pests.
- ☆ Fumigant pesticides are used on grains and crops post-harvest to reduce insects, ticks and mites, nematodes, slugs and snails and fungal diseases.

12.3.2 Handling herbicides/pesticides/fumigants

Pesticide handling

Handling pesticides is a fact of life for agricultural workers, but is part of job for anybody who cares for plants, whether in nursery, greenhouse, or in landscaping operations.

They are considered as hazardous substances since they include combination of chemicals which can cause serious harm to humans.

Safe handling of pesticides is important to farmers or farm workers.

Handling Fumigants

Farmers or farm workers are mainly exposed to fumigant pesticides by breathing them and gets exposed. Exposure to fumigant pesticides from residential structural fumigation presents a risk.

12.3.3 Using Chemicals

Chemicals on farm are dangerous which includes

- fuels,
- solvents,
- insecticides,
- herbicides,
- fungicides,
- fertilizers

Farmers need to take care when storing, transporting, using and disposing of chemicals to ensure their own safety; neighbours safety and environment. Any chemical should be treated with extreme caution and should be used as per the instructions as vapours or direct exposure can lead to many acute and chronic health effects.

Commonly used agrichemicals include:

- glyphosate
- aluminium or zinc phosphide – fumigant
- organophosphates
- pyrethroids
- methyl bromide
- sodium fluoroacetate – '1080' baits
- strychnine
- paraquat, diquat
- cresol
- alkaline and acid cleaning agents
- neonicotinoids
- formalin
- nutritional supplements – selenium and copper
- livestock vaccines and drenches

12.3.4 Hazards of agrichemicals

Since pesticides, herbicides and fumigation requires specialized agrichemicals, so farmers or farm workers should know different hazards and should monitor to see different risk level. Farmers or agricultural workers are exposed to chemicals like pesticides, veterinary drugs, solvents, and oils that causes different long-term hazards to humans.

The effects of chemical exposure to farmers or farm workers while working on fields depend on the type of chemical and degree of exposure. If chemicals are swallowed, absorbed through the skin or inhaled as mist, vapour or dust, then there appears a different immediate effect such as:

- ☆ poisoning
- ☆ headache
- ☆ nausea
- ☆ vomiting
- ☆ diarrhoea
- ☆ pinpoint pupils
- ☆ dizziness
- ☆ fine muscle twitching
- ☆ increased bronchial and lacrimal secretions
- ☆ skin rashes and irritation
- ☆ chemical burns.

Using personal protective equipment

Selecting personal protective equipment for handling hazardous chemicals such as pesticides is a complex process and should be made part of workplace safety. Farmers or farm workers need to take into consideration a variety of factors, including all identified hazards, pathways for hazardous exposure (inhalation, skin absorption, ingestion, and contact with eye or skin), and how the protective clothing will help in reducing these hazards.

During the use of pesticide, as with any hazardous chemical, farm workers must provide with:

- ☆ Correct protective clothing which should be in clean and operating condition
- ☆ Training on correct use of equipment
- ☆ Daily inspection of protective equipment for any leaks, holes, tears or wear
- ☆ Removal or repair of damaged equipment, if any
- ☆ Eye protection while harvesting, cutting, grinding to save from flying material
- ☆ Hearing protection while working with open-station tractors/implements or with loud tools as chainsaws and grinders
- ☆ Sunscreen or wide-brimmed hats to protect skin from burning and long-term risk of developing skin cancer

12.4 ASSESS RISKS IN PERFORMING JOB AND ADOPTION OF RECOMMENDED SAFE PRACTICES

12.4.1 Risk at farms

Risk is a chance that someone could be harmed by hazards like working from ladders or working around electricity. A risk assessment is an important step in protecting people, business and complying with applicable laws.

Farmers and agriculture workers are particularly exposed to ergonomic hazards such as:

- ☆ tiring or painful positions
- ☆ carrying or moving heavy loads
- ☆ standing or walking and repetitive hand or arm movements
- ☆ work-related musculoskeletal problems

Farmers live with risk and make decisions every day that affect their farming operations. Many factors that affect the decisions which farmers make could not be predicted accurately such as:

- ☆ change in weather conditions
- ☆ drop in prices at the time of harvest
- ☆ non availability of hired labour during working times
- ☆ breakage of machinery and equipment when required
- ☆ dying of draught animals
- ☆ change in government policies

The above changes are examples of risks which farmers face in managing their farm business. Such risks affect farm profitability.

12.4.2 Sources of Risk

The most common sources of risk in farming can be in areas of:

- ☆ Production: Production risk occurs from uncertainty of factors that affects quantity and quality of farm produce.
- ☆ Marketing: Marketing risk occurs due to variability of product prices and uncertainty of future market prices.
- ☆ Financial: Financial risk occurs when money is borrowed to finance the farm business.

- Institutional: Institutional risk occurs due to unpredictable changes in provision of services.
- Human: Human risk involves risk to farm business that is caused by human illness and personal situation of farmer's family.

Based on farmer's attitude, risk divides farmers among:

Risk-averse farmers

- The risk-averse farmers avoid taking risks.
- They tend to be more cautious individuals with preferences for less risky sources of income.
- They sacrifice certain amount of income to reduce the chance of low income and losses.
- They does not refuse to accept any risk at all.
- The risk averse farmer seek to be compensated for the risk taken by receiving a higher return than would normally be obtained if there were no risk.

Risk-taker farmers

- The risk-taker farmers are people who are open to more risky business options.
- They choose the alternatives which gives some chance of higher outcome, even though they may have to accept lower outcome.
- They are influenced by the return they could receive.

Risk-neutral farmers

- Risk-neutral lies between the risk-averse and risk-taking positions.

12.4.3 Workplace Risk

The main causes of risk/accidents due to equipments while working at site include:

- transport (running over or vehicle over turnings);
- falling from height (from fragile roofs, trees etc);
- struck by moving or falling objects (bales, trees etc);
- trapped by something collapsing or overturning;
- livestock related accidents and fatalities;
- asphyxiation / drowning.

Workplace Risk due to Farm vehicles

Farm vehicles and implements are involved in many farm injuries and fatalities, so number of factors that will contribute to risk, will include:

- unrestrained passengers being carried in the tray or on trailers

- unsupervised children around moving vehicles
- inexperienced drivers
- poorly maintained vehicles.

Vehicle operator should always ensure that:

- passengers use seatbelts
- the driver is trained and licensed
- adopt safe driving methods
- passengers are not carried in the back/tray
- the vehicle is maintained regularly
- children are kept away from driveways and moving vehicles.

Workplace Risk due to Harvester

The hazards due to harvester include:

- entangled with levelling or discharge of augers in grain tank
- falling from machine during pre/post season cleaning
- contacting overhead power lines
- being running over
- contacting with knife, reel, stripper rotor, straw, chopper or spreader
- being trapped under header or injured by header falling from its transport trailer

To avoid risk dont:

- climb or reach in grain tank unless the engine and all augers are stopped, and ignition key has been removed
- climb round areas of the combine unless there is purpose built access available
- carry passengers on header unless passenger seat is provided
- allow people jump on or off from harvester when it is moving
- work under an unsupported header
- carry out maintenance or try to adjust automatic sensors with the engine running
- park, carry out maintenance or extend the unloading auger when under or near over head power lines

- ☆ run the combine with guards raised or removed
- ☆ allow children on or near the combine
- ☆ brake harshly, especially on slopes.

Harvester operator should:

- ☆ make sure that anyone working on harvester has read and understood the instruction book
- ☆ carry out 'Safe stop' procedure before carrying out maintenance
- ☆ have system for thorough preseason maintenance checks to ensure that all brakes, guards and other safety devices are in effective working order
- ☆ use any grain sampling mechanism provided
- ☆ have safe system of work for combining paddocks where there are overhead power lines.
- ☆ use appropriate farm maps as reminders when daily work is being planned
- ☆ agree on routes for travel with local company which take into account safe clearance distances
- ☆ follow correct procedures for working under header or when transferring header on and off its transport trailer
- ☆ provide and use correct tools and knife guard to aid safe removal, handling and storage of the cutting knives
- ☆ work with cab door shut
- ☆ be particularly careful when reversing
- ☆ wear slip resistant footwear and avoid wearing loose clothing
- ☆ take care on steep ground and avoid sudden changes of direction.
- ☆ park across slopes if possible and ensure handbrake is fully applied before leaving the vehicle
- ☆ make sure all guards are in position and correctly fitted before starting work
- ☆ regularly clean straw and chaff deposits from the engine compartment and around belts and pulleys
- ☆ make sure there is suitable fire-fighting equipment available on the combine

General safety

- ☆ Make sure all guards and shields are in place.
- ☆ Ensure all workers are trained in proper operating technique and safety procedures when using machinery.

- ☆ Stop all moving parts before servicing the machine.
- ☆ Operators should take regular breaks.
- ☆ Protective footwear and close-fitting clothes are essential when working in and around machinery.
- ☆ Use suitable access equipment and take precautions for work at height during cleaning and maintenance operations.
- ☆ Be patient when working in difficult crop conditions such as laid crops or unfamiliar fibrous crops like linseed.
- ☆ Don't operate the machine beyond its capacity.
- ☆ Use your instruments and watch and listen for potential blockages – avoiding blockages is easier than clearing them.
- ☆ Use drive reversing mechanisms, when fitted, to wind out blockages.
- ☆ Remember the baler operator – try to avoid leaving heaps in the swath.
- ☆ Remember the hazards posed by straw choppers and spreaders – allow adequate rundown time before approaching the rear of the combine.
- ☆ Combines fitted with yield measuring meters using an ionising radiation source need additional safety procedures which must be laid down, clearly understood and implemented.
- ☆ Exposure to high levels of grain dust causes ill health, including occupational asthma, farmer's lung, grain fever, chronic bronchitis, allergic eye and nasal infections.
- ☆ Make sure the cab filter is of the type recommended by the manufacturer and is fitted and maintained in accordance with the manufacturer's instructions.
- ☆ When unloading on the move consider the other driver and avoid sudden manoeuvres.

12.4.4 Managing Risk

Reducing risk are often used in combination with one another as no single strategy can cover all of risk likely to be encountered. Farmers need to consider the risks simultaneously and to develop an integrated approach for better management. They need to recognize the advantages and disadvantages of each risk management option both individually and in combination. Each farmers should select an appropriate strategy based on their goals, attitudes towards risk and their personal and financial situations.

Farmers should choose and combine strategies based on their goals, attitudes towards risk and their personal and financial situations. Every farmer family required to find their own ways of coping with risk. Farmers' responses to risk are as diverse as the risks that affect their farms.

Farmers can reduce risk by learning about and applying new technologies and practices designed to address specific risks common to their area of production such as new varieties of seed are being developed and livestock are being bred with certain characteristics like:

- drought-resistant seed for wheat;
- bird-resistant seed for sorghum;
- disease- and pest-resistant seed species;
- disease-resistant livestock species;
- livestock bred to provide better productivity;
- irrigation for high-value crops;
- crops and livestock bred specifically to improve marketability.

Factors involved in managing risk

Avoiding Production Risk

- Selecting low-risk activities is a way to reduce production risk where farmers choose reliability over potential profitability.
- In this, farmer may forego an enterprise that has a high potential for income but also carries high risk for loss, and choose instead an enterprise which is less profitable but also less risky.

Using Flexible Farming

- Farming system flexibility is an important strategy for managing risk.
- A flexible farming system makes it possible for farmer to make quick or short-term changes in production and sales.
- Farmers who sell cash crops also reduce risk by using available funds to enable them to change to other enterprise if price of main cash crop falls.
- By keeping farm systems flexible, farmers are able to make decisions in response to changing circumstances.
- If farmers work on general production plans, they should keep their options as open as possible in order to respond to opportunities and risks as they occur.

Production diversification

- ☆ Diversification spreads risk and is successful risk management strategy as not all farm enterprises and operations are likely to be affected in the same way by changing situations. It involves:
- ☆ Managing multiple farm enterprises together at any one time (or in the same season);
- ☆ Engaging in the same farm enterprise in different physical locations;
- ☆ Engaging in the same farm enterprise over successive periods of time (or seasons);
- ☆ Generating income from off-farm activities.

Engaging in same enterprise in different physical locations

- ☆ Such type of risk managing strategy works on understanding that the same crops grown in different areas will not meet the same fate.
- ☆ Perhaps a crop grown in the area surrounding the house is more likely to be infested by rodents than the same crop grown elsewhere.

12.5 DISPOSE OF WASTE MATERIAL SAFELY AND CORRECTLY IN A DESIGNED AREA

12.5.1 Agriculture waste

Agricultural waste is defined as unwanted waste produced as a result of agricultural activities such as:

- ☆ manure,
- ☆ oil,
- ☆ silage plastics,
- ☆ fertilizer,
- ☆ pesticides and herbicides
- ☆ farms and poultry
- ☆ veterinary medicines
- ☆ horticultural plastics

Waste poses a threat to public health and environment if not stored, collected and disposed of properly. The perception of waste as an unwanted material with no intrinsic value has dominated attitudes towards disposal. Dispose of farm waste should be done in accordance with local regulations; where possible using specialised hazardous waste disposal systems.

If wastes are not properly handled, they can pollute surface and groundwater and contribute to air pollution. The proper management of waste from agricultural operations can contribute in a significant way to farm operations. Waste management helps to maintain a healthy environment for farm animals and can reduce the need for commercial fertilizers while providing other nutrients needed for crop production. The waste which is reduce, recycle and make it usable for different purpose is a waste management.

Fig Different Farm waste

Farm waste must be managed with caution prior to disposal, for this, farm workers should safely store the farm waste in a closed and possibly locked area or container to ensure safety for family, livestock, pets and wildlife. Farm workers or farmers should carefully dispose of unwanted or expired pesticides as pesticides are hazardous wastes and cannot be disposed off in sanitary landfills or by burning.

On-farm dumps, though generally not recommended, can be used exclusively for disposal of small amounts of inert materials. The dump should be located on a naturally dry site, and be fenced to prevent entry by children and animals.

12.5.2 Disposal of waste

The disposal options can be:

- ☆ Transportation to a rendering plant for disposal.
- ☆ Burial in farm pit.
- ☆ Burning within regulations.
- ☆ Composting.
- ☆ Sending to a Class I or II landfill.
- ☆ Natural disposal (except for animals that have been euthanized with drugs and chemicals).

In the absence of the facilities to do this, the following procedures for disposal may be used as a last resort:

- ☆ Metal cans and drums: wash out, puncture, crush and bury
- ☆ Plastic: wash out, puncture and bury
- ☆ Cardboard packaging: burn or bury.

If farm wastes, including deadstock, are managed inappropriately, they can cause contamination of the premises, livestock, inputs and food. The accumulation of farm waste can also provide habitats and food for pests.

12.5.3 Safe disposal procedures for agrichemicals

Suggestions for safe disposal of chemicals include:

- ☆ Always follow the manufacturer's instructions for proper disposal of both chemicals and reinstate from equipment.
- ☆ Thoroughly triple rinse and then puncture empty containers to prevent reuse for other purposes.
- ☆ Return empty containers to the manufacturer or check with your local council about proper disposal methods.

- ☆ Audit your chemical store on a regular basis and dispose of any excess or outdated chemicals in the appropriate manner.

12.5.4 Waste Treatment process

Many waste treatment processes should be performed like:

- ☆ When dealing with agricultural waste, farmers and farm workers should follow health and safety regulations.
- ☆ Farmers and farm workers should provide written instructions for storing and disposing of each type of waste.
- ☆ Farmers and farm workers should dispose waste if they determined that they will not use prevention, preparation for reuse, recycling or any other recovery method.

Agriculture waste treatment process includes:

Composting

- ☆ It is a method where organic matter present in agricultural waste gets decomposed by aerobically/anaerobically using biochemical process which gets converted into humus.
- ☆ It gets divided into preparing agricultural waste, decomposition, product preparation & marketing.
- ☆ Preparation of agricultural waste shall be:
    - free of material that is not produced in agricultural field.
    - reasonably free of dirt, soil and visible surface.
    - arranged so that it will burn with a minimum of smoke.

Decomposition

- ☆ Waste is decomposed using Nadep system, Vermi compost decomposition, Anaerobically decomposition

Fig Nadep process

Fig Vermi compost decomposition

Landfill:

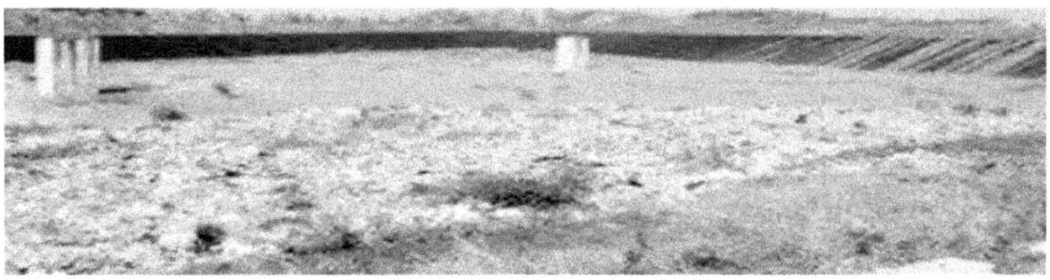

Fig Sanitary landfill

- A landfill site (rubbish dump or dumping ground) is a site for disposal of waste materials by burial
- It serves as the most common method of dumping waste.
- Many landfills are used for waste management purposes like sorting, treatment, or recycling
- In sanitary landfill, waste gets compacted and covered with soil and if disposal site has reached its capacity, then a final layer of 2ft is applied

12.6 INSTRUCTIONS TO MINIMIZE ENVIRONMENTAL DAMAGE

12.6.1 Adopting different farming practices

Farmers work hard to produce food to feed, and also to protect the environment for next generation. There are many different practices which farmers should adopt in order to ensure, that they are protecting the environment using different farming practices such as:

- conservation
- preservation
- moderation

Conservation:

- ☆ It shows use of less resources so that it will have less effect on the land.
- ☆ Farmers uses different practices like no-till, putting fencing, planting cover crops, tractor collecting water runoff, integrating crop and pasture rotations.
- ☆ No-till, or zero tillage is a practice where farmers do not use tillage methods as using this, they do not disturb the soil until the seed is planted.
- ☆ Collecting runoff water from fields is an important way that farmers help to stop pollutants from reaching water sources.
- ☆ For cleaning of water, farmers uses fences around the streams, rivers and ponds to keep livestock out of water.
- ☆ Planting of cover crop during winters where other plants cannot grow as it will help to stop soil erosion.

Preservation:

- ☆ Preservation helps to keep things the way they are, to keep them for future generations and to improve their quality.
- ☆ Farmers can practice preservation in various ways.

Moderation

- ☆ Using less resources and developing new ideas to manage resources on farm become a way for farmers to help protecting the land.
- ☆ Farmers uses moderation to apply nutrients to land.
- ☆ Farmers realize that putting too much pesticide, insecticide, or fertilizer on a field does not help the crops, but instead harms the environment.
- ☆ New methods of application have been developed for farmers to decrease their impact on the land.

12.6.2 Adopting different methods to reduce environmental impact

Farmers should adopt new methods which will help in reducing environmental impact:

- ☆ Mineral fertilizer: Use of environment friendly mineral fertiliser which are produced with the help of technology reducing greenhouse gas emissions. Production, transportation and use of mineral fertilisers contribute directly and indirectly to emissions of greenhouse gases, carbon dioxide (CO_2) and nitrous oxide (N_2O). Such fertilisers enhance agricultural productivity and stimulate CO_2 uptake by the crop. They tends to increase yield and reduce necessity to cultivate new land. In such fertiliser's life cycle, emissions and uptakes of greenhouse gases occur.

- ☆ Precision cultivation: In this, the nutrient gets added only where they are needed by growing crop. Phosphorus (P) and nitrogen (N) that are not absorbed by plants leak into lakes, rivers and oceans, causing eutrophication. Such cultivation is beneficial in climate perspective, and serves the financial interests of farmers. Use of such farming will allow fertiliser amount and distribution are tailored to actual needs.
- ☆ Using green energy which is produced using electricity with guarantees of origin.
- ☆ Use of ecodriving as tractors, farm machines and other vehicles uses energy efficiently where operators are trained in ecodriving.
- ☆ Adopting energy monitoring at facilities, since consumption is analyzed and measures are taken to maximize energy efficiency.

12.6.3 Recycling of Waste

- ☆ It is a process to change waste into new products.
- ☆ It prevent waste of potentially useful materials and reduces consumption of fresh raw materials and energy usage.
- ☆ It reduce air pollution from incineration and water pollution from land filling.
- ☆ It is the main unit of modern waste reduction and is considered as third component of Reduce, Reuse, Recycle.

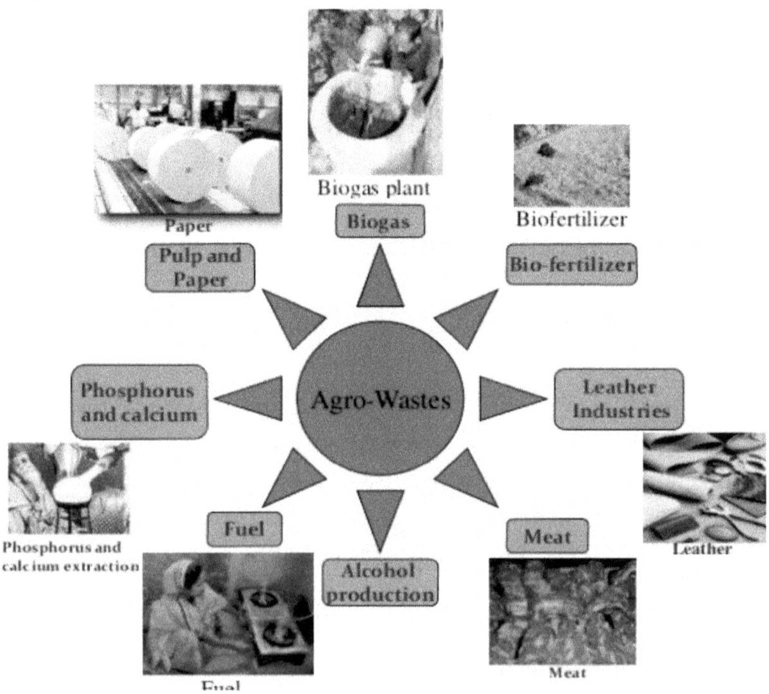

12.7 REPORTING OF MISHAPPENING AT SITE

In case of any mishappening at site, site workers should report the incidence to farm manager who further will records all the activity in a register for further consultation.

12.7.1 Types of reportable incidents

In farms there are many incidences that take place which could be:

- ☆ Accidents due to farming vehicles and machinery
- ☆ Falling off from or through fragile roofs
- ☆ In case if someone dies or became unconscious
- ☆ Someone injured due to work-related accident
- ☆ If some worker commit or tend to commit suicides
- ☆ Physical violence among workers
- ☆ Specific burn related injuries
- ☆ Fracture due to falling or while working
- ☆ Blood coming from fingers, thumbs and toes
- ☆ Injury due to use of toxic chemical at sight
- ☆ Any loss of consciousness caused by head injury or asphyxia
- ☆ Any injury arising from working in an enclosed space which leads to hypothermia or heat-induced illness
- ☆ Collapse, overturning or failure of load-bearing parts of equipment
- ☆ Plant or equipment which comes in contact with overhead power lines
- ☆ Leakage of gas, fumes
- ☆ Danger due to design, construction, installation, modification or servicing of appliance or fitting

12.7.2 Action Taken while seeing Incidence

After seeing an incidence at site, the site workers/supervisor should immediately give first aid to the injured person. It is the duty of farm workers or supervisor to report any mishapenning at site to their superiors. The supervisor should note the type of mishapenning and should take action based on details collected.

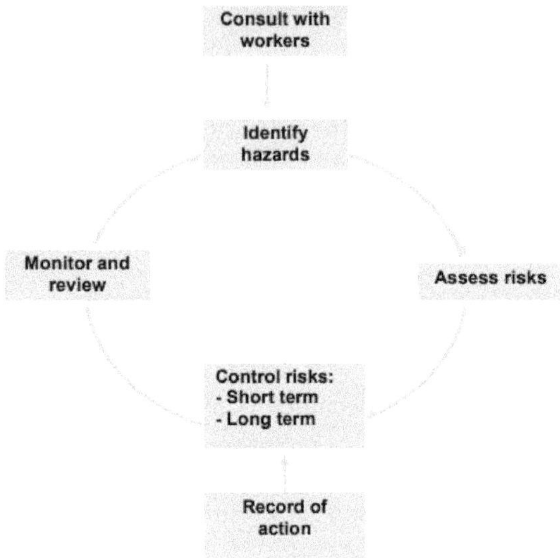

The supervisor should note and discuss the problem and should involve in monitoring of the mishapenning that takes place at site.

12.8 FIRST AID EQUIPMENTS/ ARRANGE AND STORE FIRST AID EQUIPMENTS PROPERLY

Accidents on farms and ranches can be quite severe, and space in a first aid kit is limited, so it is important to choose items for kits wisely. Most farms and ranches require multiple first aid kits due to the many types of jobs and the dispersed areas of work in a production agriculture operation. Not only is it important to have appropriate first aid kits on your farm or ranch, it is important that you and others in your operation understand basic first aid and cardiopulmonary resuscitation (CPR).

12.8.1 Items in first aid kit

Injuries vary from job to job in production agriculture, so first aid kits should be tailored to the potential injury that could result from a particular job. Immediately and properly examining and treating of injuries save life and is essential to reduce pain and help injured people make a quick recovery.

The following items should be there in first aid kit which includes:

Sterile first aid dressings of different sizes

- ☆ Trauma dressings for covering large areas
- ☆ Small, sterile adhesive compresses in sealed envelopes
- ☆ Roller bandages with different sized cling bandages

- ☆ Rolls of adhesive tape in assorted widths
- ☆ Triangle bandages to use as slings/coverings over large dressings
- ☆ Antiseptic wash
- ☆ Tongue depressors
- ☆ Bandage scissors and heavy-duty scissors to cut cloth
- ☆ Tweezers to remove insect stingers or small splinters
- ☆ Splints of required size for splinting broken arms and legs
- ☆ Sterile saline solution
- ☆ Safety pins
- ☆ Chemical ice bags to reduce swelling
- ☆ A pocket mask for resuscitation
- ☆ Disposable rubber gloves
- ☆ Eye goggles
- ☆ Emergency blanket

It is the duty of medical supervisor/farm manager to check the first aid kit on regular basis. Medications expire, so you may need to replace old ones. You need to replenish supplies that you have used in the past, so make sure that more common treatments such as medications, bandages etc are always available. If possible use antiseptic solution to clean the scissors and injections.

INDEX

www.ingramcontent.com/pod-product-compliance
Ingram Content Group UK Ltd.
Pitfield, Milton Keynes, MK11 3LW, UK
UKHW021328190726
13845UKWH00004B/68

9 789349 964037